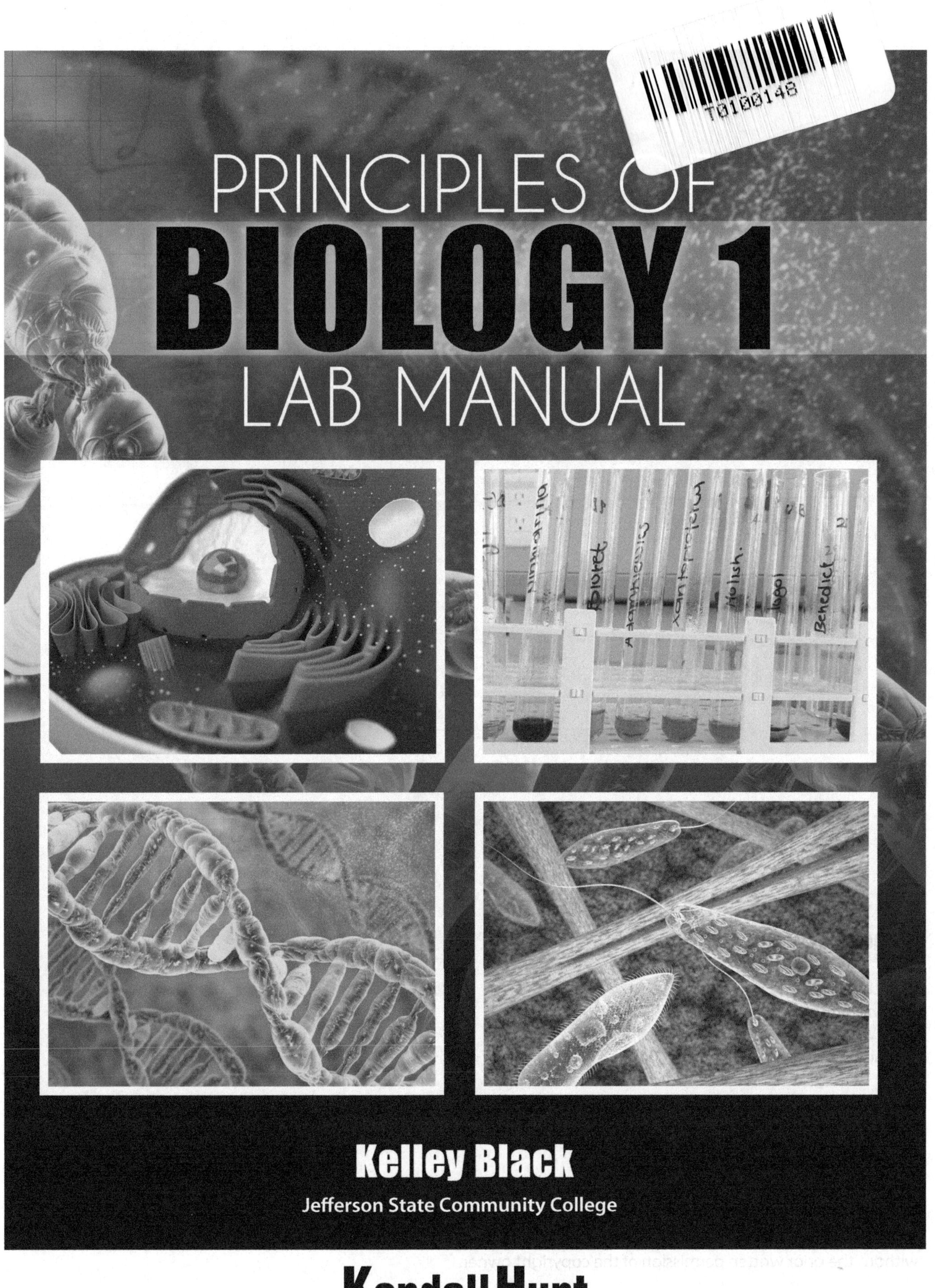

Kendall Hunt
publishing company

Kendall Hunt
publishing company

www.kendallhunt.com
Send all inquiries to:
4050 Westmark Drive
Dubuque, IA 52004-1840

ISBN 978-1-5249-8425-0

Published in the United States of America

Contents

Objectives

To be able to

1. Understand the metric system and be able to convert from one metric unit to another.
2. Know the English equivalents between: meter and yard; centimeter and inch; liter and quart; kilogram and pound.
3. Recognize the following pieces of equipment:

meter stick	Bunsen burner lighter
beakers	Bunsen burner
Erlenmeyer flasks	tripod
graduated cylinders	wire gauze
test tubes	spot plate
test tube racks	thermometers
centrifuge tubes	balance

4. Describe the proper use and function of the equipment listed above.
5. Demonstrate ability to measure length, volume, and mass properly.
6. Demonstrate ability to light a Bunsen burner.
7. Know the following temperatures in °F and °C: human body temperature; freezing and boiling points of water; room temperature.
8. Complete the Laboratory Report Sheet.

Introduction

There are two main purposes to this lab. One is to learn the metric system and the other is to become familiar with some of the laboratory equipment used in a biology lab. The metric system is used in science because it is universal. It is a decimal system based on units of ten. To convert from one unit to another, the decimal point is moved to the right or to the left. Figure 1.1 lists the common metric prefixes along with some of the English conversions. The conversion from one metric unit to another will be required. The basic metric unit of length is the meter; the basic unit of volume is the liter; and the basic unit of mass is the gram.

Metric Prefixes

Multiplication Factor		*Prefix*	*Symbol*
1000	$= 10^3$	kilo	k
0.1	$= 10^{-1}$	deci	d
0.01	$= 10^{-2}$	centi	c
0.001	$= 10^{-3}$	milli	m
0.000001	$= 10^{-6}$	micro	μ
0.000000001	$= 10^{-9}$	nano	n
0.0000000001	$= 10^{-10}$	angstrom	Å

	Metric System	*English System*
Length	1 meter (m)	= 39.37 inches = 3.28 ft.
	1 centimeter (cm) = 1/100 m	= 0.39 inches
	1 millimeter (mm) = 1/1,000 m = 1/10 cm	= 0.039 in. = 1/25 in.
	1 micrometer (μm) = 1/1,000 mm = 1/10,000 cm	= 0.000039 in. = 1/25,000 in.
	1 nanometer (nm) = 1/1.000 μm	= 1/25,000,000 in.
	1 angstrom unit (Å) = 1/10 nm	= 1/250,000,000 in.
Mass (weight)	1 kilogram (kg) = 1,000 g	= 2.205 pounds
	1 gram (g)	= 0.03527 ounces
	1 milligram (mg) = 1/1,000 g	= 0.00003527 ounce
Volume	1 liter (l)	= 1.06 qts.
	1 milliliter (ml) = 1/1,000 liter	= 0.00106 qts.

Figure 1.1

Procedure

A. Metric System

1. Refer to Figure 1.2 and study the metric prefixes.
2. Observe the meter stick provided. Notice the English equivalent on the opposite side. On the metric side, observe the ten major divisions. Each division is a decimeter, which represents 0.1 or 1/10 of the meter. Therefore, there are 10 decimeters in one meter. Observe the 100 divisions, each representing a centimeter or 0.01 or 1/100 m. The small black lines on the meter stick represent the millimeters, each being .001 or 1/1000 meter. The other prefixes in Figure 1.2 are too small to see on a meter stick.
3. Refer to Figure 1.2, which illustrates the steps in converting between the metric units. The column on the far right indicates the number of decimal place differences between the prefixes. If the conversion is from a larger unit to a smaller unit, the decimal point is moved to the right. If the conversion is from a smaller unit to a larger unit, the decimal point is moved to the left.

Look at the example in Figure 1.2. Using this information, complete number 1 on the Laboratory Report Sheet.

4. The metric system uses the Celsius or centigrade scale ranging from 0 to 100 degrees. Study the temperatures listed in Figure 1.3.

Metric Conversions

The DRUL rule

D own
R ight
U p
L eft

Prefix	Power
Kilo (k)	3
	2
	1
Basic unit—Gram (g), Liter (l), Meter (m)	**0**
	−1
Centi (c)	−2
Milli (m)	−3
	−4
	−5
Micro (μ)	−6
	−7
	−8
Nano (n)	−9
Angstrom (Å)	−10

When converting from one metric unit to another, answer the following three questions.

1. Does the conversion go up or down?
2. Does the decimal point go to the right or left?
3. How many places does it move?

Examples:

a. 25.0 mm = ? nm

1. Does the conversion go up or down? (mm to nm **down**)
2. Does the decimal point move right or left? (Down Right—**right**)
3. How many places? **(6)**

Therefore, 25.0 mm = 25000000.0 nm

b. 2.5 μm = ? cm

1. Does the conversion go up or down? μm to cm is **up**
2. Does the decimal point move right or left? (UP left—**left**)
3. How many places? **(4)**

Therefore, 2.5 μm = 0.00025 cm

Figure 1.2

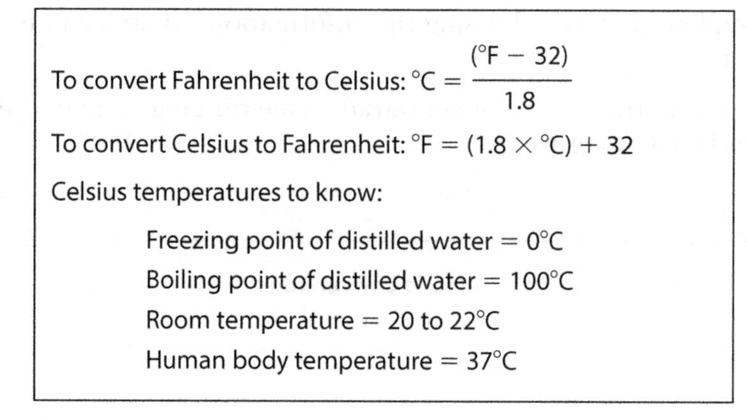

To convert Fahrenheit to Celsius: $°C = \frac{(°F - 32)}{1.8}$

To convert Celsius to Fahrenheit: $°F = (1.8 \times °C) + 32$

Celsius temperatures to know:

Freezing point of distilled water = 0°C
Boiling point of distilled water = 100°C
Room temperature = 20 to 22°C
Human body temperature = 37°C

Figure 1.3

B. Laboratory Equipment

1. Identify the lab equipment that has been provided.
2. Observe the glassware used for measuring volumes of liquid: beakers, Erlenmeyer flasks, and graduated cylinders. Note the ml markings on each and the variety of sizes. When measuring the volume of a liquid, the beakers and flasks are not as accurate as the graduated cylinders. Use the size of the graduated cylinder that is most appropriate for the volume being measured. To measure 5 ml, a 10 ml graduate is more accurate than a 100 ml graduated cylinder. Hold the graduated cylinder at eye level and notice that the upper surface of the liquid is curved. This curve is called the meniscus. The bottom of the meniscus should be at the desired volume. Beakers and flasks are often used for approximate amounts and to hold reagents taken from a stock bottle. **Never return any unused reagent to a stock bottle, and do not place a pipette or dropper into the stock bottle.**
3. Observe two types of pipettes: the graduated pipette and the Pasteur pipette. The graduated pipette is used to accurately measure smaller volumes of liquid. The Pasteur pipette is not calibrated and is used to dispense drops of liquid.
4. Test tubes can be used for testing solutions, among other purposes. If a tube is to be heated, make sure it is heat resistant. The test tube holder is used to handle hot test tubes. Do not squeeze the holder when carrying the test tube. When heating a solution in a test tube, do not point the opening toward anyone. A test tube rack is used to hold multiple test tubes in an upright position.
5. Distinguish between the test tube and the centrifuge tube. The centrifuge tube is pointed at the bottom and is used in a centrifuge to separate materials according to density.
6. The Bunsen burner may be used as a source of heat. The lab instructor will demonstrate its proper use. Securely connect the tubing of the Bunsen burner to the gas jet. Turn the valve at the base of the burner counter-clockwise to stop the flow of gas. Turn on the gas outlet. Open the valve at the base of the burner to start the flow of gas. Using the Bunsen burner lighter, strike the flint until a flame appears. Adjust the valve of the burner until a cone-shaped blue flame is seen. **Keep hair and clothing away from the flame. Always turn off the gas outlet when done with the burner.**

7. The three-legged tripod and wire gauze can be used to place a beaker on while heating with the Bunsen burner.
8. Observe the spot plates. Note the depressions, which are used to test materials. For example, to show the test for starch, a few drops of starch solution are placed in the depression. If iodine is added to it, a black or purple color results, indicating a positive test.
9. Carefully inspect the thermometers, noting both the Fahrenheit and centigrade scales.
10. The triple beam balance is used for weighing. Handle them with care. Before weighing, the sliders are moved along the beam until the indicator points to zero. This is called zeroing the balance. Make sure that the sliders are securely in the grooves. The object to be weighed is placed on the pan and the process repeated. Weights can be recorded to the nearest tenth of a gram.
11. The proper use of the equipment can be demonstrated by answering the questions on the Laboratory Report Sheet.

Laboratory Report Sheet

NAME ______________________ **SECTION** __________ **GRADE** __________

Introduction to the Laboratory

1. Complete the following conversions:

   a. 3.2 km = ______________ m

   b. 0.015 µm = ______________ nm

   c. 27.83 m = ______________ cm

   d. 5000 ml = ______________ l

   e. 0.95 mg = ______________ g

   f. 6513 Å = ______________ nm

   g. 4.92 kg = ______________ g

   h. 13.2 dm = ______________ mm

   i. 0.002 mm = ______________ m

   j. 98.6 °F = ______________ °C

2. Using the meter stick, measure the length of the lab bench in:

   __________ m __________ cm __________ mm __________ µm

3. Determine your height in:

   __________ m __________ cm

4. Select the proper piece of glassware to accurately measure the designated volume of water:

   3.5 ml ______________

   750 ml ______________

   82 ml ______________

   Measure 7.0 ml of water and check the accuracy with the instructor.

5. Weigh a 500 ml beaker on the balance provided to the nearest tenth of a gram.

   __________ g

6. Measure the temperature of the tap water and check the accuracy with the instructor.

   __________ °C __________ °F

2 pH AND BUFFERS

Objectives

To be able to

1. Define and give examples of acids.
2. Define and give examples of bases.
3. Use laboratory equipment to measure pH of a sample.
4. Write and interpret a neutralization reaction.
5. Explain how a buffer works and demonstrate with written chemical equations.
6. Explain the importance and functioning of the carbonic acid–bicarbonate buffer system in the human body.

Terms to Define

1. Acid ______________________________
2. Base ______________________________
3. Buffer ______________________________
4. pH indicator ______________________________

Acids, Bases, and pH

Molecules that form from ionic bonds can lose or gain electrons to form ions. Ions are charged particles and are also known as electrolytes because they conduct electricity. An acid is a molecule that releases hydrogen ions (H^+) when dissolved in water. A hydrogen ion is a hydrogen atom that has lost its electron so it has a positive charge. When a molecule's ions separate from each other, this is known as dissociation. A solution with a high concentration of H^+ is termed an acid. Bases, on the other hand, bind to and remove H^+ from solutions. A base or alkaline substance contains hydroxide ions (OH^-) which bind to H^+ to form water (H_2O). It is important to know the number of hydrogen ions in a solution, and this is known as the pH. The pH scale runs from 0 to 14. A pH value of 7 is considered neutral which means that the number of $H^+ = OH^-$. If the concentration of H^+ increases, the pH value decreases (this is known as an inverse relationship, see Figure 2.1). Each value on the pH scale reflects a ten-fold change in the H^+ concentration. For example, a solution with a pH of 2 has 10× more H^+ than a solution with a pH of 3 and 100× more H^+ than a solution with a pH of 4. Conversely, the higher the pH value, the fewer H^+ in the solution. A solution with a pH of 10 has 10× fewer H^+ than a solution with a pH of 9. Small changes in the pH unit reflect large changes in the concentration of H^+. The pH of human arterial blood is kept between 7.35 and 7.45 and the pH limits for human life are between 7.0 and 7.8!

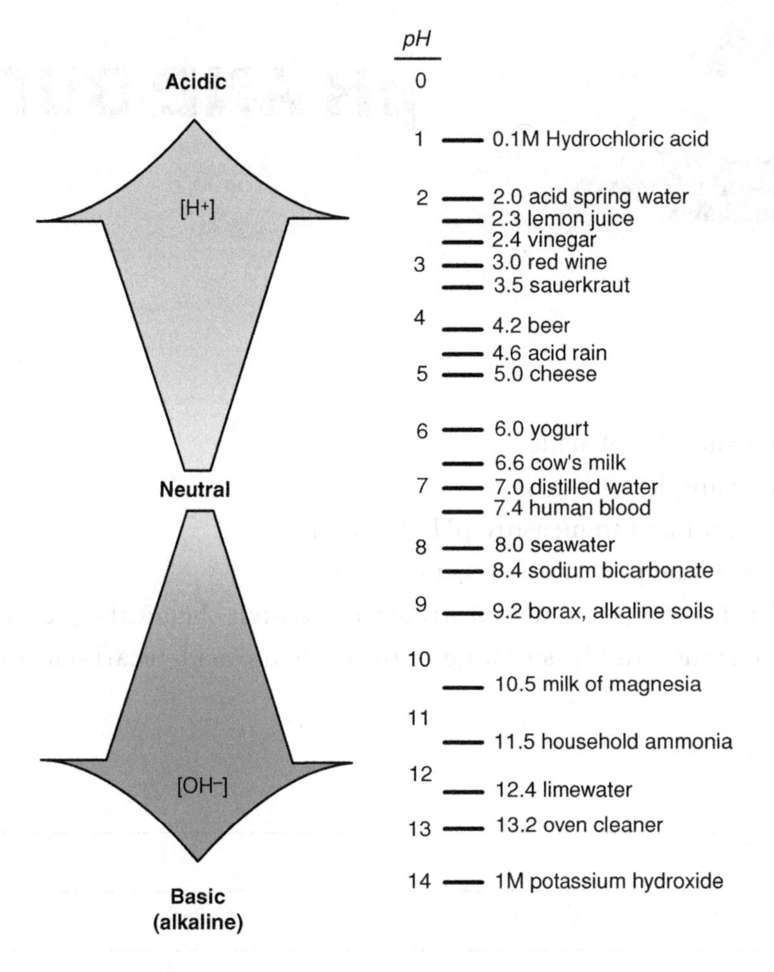

Figure 2.1 Notice the inverse relationship between the pH value and H^+ concentration. As the H^+ concentration increases, the pH value decreases.

ACTIVITY 1: Testing pH with pH Indicators, Test Strips, and a pH Meter

There are several methods to measure pH. There are advantages and disadvantages to each. You will measure the pH of 5 solutions (A to E) using different techniques. The techniques you will use are: the pH indicators Bromthy-mol blue and phenolphthalein, pH paper, and a pH meter.

Procedure

1. Obtain the laboratory bin with your solutions (labeled A to E).
2. Label two test tubes each for solutions A, B, C, D, and E.
3. Place 10 mL of solution A into **one** of its labeled tubes.
4. Use a pair of forceps to dip a piece of pHydrion paper into solution A.
5. Compare the color of the pHydrion paper with the information on the paper dispenser chart and record the result in Table 2.1.

6. Use the pH meter to test the pH of solution A.

DIGITAL pH METER INSTRUCTIONS

1. Remove the black cap from the tip of the pH meter. Place it somewhere safe, so that it is not lost.
2. Use a squirt bottle containing deionized water to rinse the pH meter probe over a waste container.
3. Gently dry the tip of the pH meter with a Kimwipe®.
4. Lower the pH meter into the liquid and allow the reading to stabilize. Record the pH.
5. Remove the pH meter and rinse the tip of the pH meter with deionized water over a waste container.
6. Gently dry the tip of the pH meter with a Kimwipe®.
7. Record the next solution that you wish to measure by repeating steps 4 to 7.

7. Pour half of the solution from tube into the second solution A test tube. One set of test tubes will undergo pH testing using the pH indicator bromthymol blue, and the other set of test tubes will undergo pH testing using the pH indicator phenolphthalein. (NOTE: a pH indicator is a substance that changes color to indicate if the solution is acidic or alkaline.)
8. Add a few drops of bromthymol blue to the first test tube, and record the color for solution A in Table 2.1.
9. Add a few drops of phenolphthalein into the second solution A test tube, and record the color in Table 2.1.
10. Repeat the procedures with solutions B, C, D, and E and record results in Table 2.1.

Be careful to avoid cross-contamination of test equipment and solutions!

Table 2.1 **Results for pH Testing of Solutions A–E**

Solution	pH from pH Meter	pH from pH Paper	Color after Bromthymol Blue	Color after Phenolphthalein
A				
B				
C				
D				
E				

Answer the following questions:

1. What color does bromthymol blue turn when mixed with an acid? ____________________
2. What color is it when mixed in a base? ____________________
3. What color is it when in a neutral solution? ____________________

4. What color does phenolphthalein become when mixed with a base? ____________________

5. What color does phenolphthalein become when mixed in an acid? ____________________

6. Which pH method do you think is the most accurate? ____________________

7. What is one advantage to using pH paper rather than the pH indicators?

____

8. What is one advantage to using pH meter rather than the pH paper?

____

ACTIVITY 2: Identifying pH of Common Solutions

Now that you know how to test pH, we will measure the pH of common household solutions. You will get a small sample of each solution from the supply area in lab. First predict what pH you think each will have, and then choose which method you will use to measure the pH of the solution. Were your predictions correct?

Solution	Predicted pH	Measured pH
Bleach		
Vinegar		
Lemon juice		
Milk		
Cola		
Coffee		
Mineral water		
Tap water		
Laundry detergent		
Maalox®		

Answer the following questions:

1. What pH do cleaning agents have? ____________________

2. Which solution was the most acidic? ____________________

3. Which solution was the most alkaline? ____________________

Buffers

It is critical for the body's homeostasis that pH be maintained. Disruptions in pH can disrupt protein function and ultimately depress the central nervous system (CNS) to cause coma and death. A buffer is a molecule that maintains pH by either releasing H^+ if the pH is too high or removing them if the pH falls. Buffers are usually in pairs consisting of a weak acid (substance that can donate H^+ but does not completely dissociate) and a weak base (substance that can bind to and remove H^+ ions). Therefore, a good buffer does not experience rapid pH changes. The buffering ability of a solution can be tested by adding an acid or a base and checking the pH after each addition. If the pH changes slightly it is a good buffer. Buffers can reach their capacity and this is seen when the pH changes dramatically after the addition of the acid or base.

Note that when an acid is mixed with a base, a salt and water are formed. This is termed a neutralization reaction and is shown here.

$$\underset{\text{Hydrochloric acid}}{HCl} + \underset{\text{Sodium hydroxide}}{NaOH} \rightarrow \underset{\text{Sodium chloride}}{NaCl} + \underset{\text{Water}}{H_2O}$$

There are two important buffer systems in the human body: the phosphate buffer system and the carbonic acid–bicarbonate buffer system. Both utilize a weak acid and a weak base to maintain pH in response to acidic or alkaline conditions. The phosphate buffer functions in the kidney to buffer urine, and the carbonic acid–bicarbonate buffer is the major buffer in the blood.

Phosphate Buffer: NaH_2PO_4 (sodium dihydrogen phosphate) is the weak acid and donates H^+ if pH increases.

Na_2HPO_4 (sodium monohydrogen phosphate) is the weak base and picks up H^+ if pH decreases.

So, if a strong acid was added to the solution, the weak base of the buffer pair will buffer the strong acid by converting it into a weak acid which prevents a pH change. The buffer replaces the strong acid with a weak acid.

$$\underset{\text{(strong acid)}}{HCl} + \underset{\text{(weak base)}}{Na_2HPO_4} \rightarrow \underset{\text{(weak acid)}}{NaH_2PO_4} + \underset{\text{(salt)}}{NaCl}$$

And if a strong base needs to be buffered, the weak acid of the buffer pair will combine with the strong base to convert it into a weak base to prevent a pH change.

$$\underset{\text{(strong base)}}{NaOH} + \underset{\text{(weak acid)}}{NaH_2PO_4} \rightarrow \underset{\text{(weak base)}}{Na_2HPO_4} + \underset{\text{(water)}}{HOH\ (H_2O)}$$

Carbonic Acid–Bicarbonate Buffer: H_2CO_3 (carbonic acid) is the weak acid and donates H^+ if pH increases.

HCO_3^- (bicarbonate ion) is the weak base and picks up H^+ if pH decreases.

This system relies on the kidneys and the lungs for buffering. In fact, these are the two most important organs in the human body for maintaining pH and homeostasis. The kidneys control the amount of HCO_3 while the lungs control the amount of CO_2. The equation is shown here.

$$CO_2 + H_2O \leftrightarrow H_2CO_3 \leftrightarrow HCO_3^- + H^+$$

If the H^+ concentration increases, the addition of H^+ pushes the equation to the left. Excess H^+ are picked up by HCO_3^-, and more H_2CO_3 is formed which travels to the lungs where it is broken down into water and CO_2 which is released in exhaled air. This results in an increased respiratory rate (RR).

If, on the other hand, the H^+ concentration decreases, more H^+ must be generated. This is done by retaining CO_2 (by slow, shallow breathing) so it can combine with H_2O to form more H_2CO_3. The H_2CO_3 releases H^+ and restores pH.

We will assess some common solutions and see if they are good chemical buffers by measuring the pH change as acids and then bases are added to them. We will use the pH meter to measure the pH change and record data in the table provided.

ACTIVITY 3: Assessing the Buffering Capacity of Distilled Water

Procedure

1. Place 50 mL of distilled water in a beaker, and place it on the stirrer station. Secure the pH meter with the clamp above the beaker. Put the CLEAN magnetic stir bar in the beaker.
2. Turn the dial on the stir station slowly until the stir bar spins at the desired speed. The stir bar should not spin so fast that the liquid spills out of the beaker.
3. Lower the pH meter into the liquid and allow the reading to stabilize. Record the pH.
4. After you have recorded the starting pH, keep the pH probe in the solution.
5. Add 2 mL of acid to the solution. Allow a short time for the acid to mix thoroughly in to the solution.
6. Allow the pH meter reading to stabilize. Record the pH.
7. Repeat steps 5 and 6 until you have finished making all of your measurements needed for that solution. *You will test pH every time you add 2 mL of the acid (0.1 molar HCl) until you reach 12 mL.*
8. To start measuring a new solution, raise the pH meter and rinse the probe with deionized water over a waste container and dry the probe with a Kimwipe®.
9. Place the new solution with a stir bar on the stir station, and begin recording pH again.

Repeat the procedure using a fresh supply of distilled water and a clean beaker. Make sure you thoroughly rinse the probe tip. This time you will add 2 mL quantities of a base (0.1 molar NaOH) for a total of 12 mL.

mL of HCl Added to Distilled Water	pH
0 (initial pH)	
2	
4	
6	
8	
10	
12	

mL of NaOH Added to Distilled Water	pH
0 (initial pH)	
2	
4	
6	
8	
10	
12	

Activity 4: Assessing Buffer Ability of Other Solutions

Next you will assess the buffering ability of some solutions commonly used to treat a "sour stomach." Let's see how they measure up for buffering ability for an acid. Each lab group will test the effectiveness of a different solution. The solutions are in your bin.

Group 1: Chemical buffer
Group 2: Coca-Cola
Group 3: Alka-Seltzer
Group 4: Ginger ale
Group 5: Milk
Group 6: Baking soda

You will repeat the procedure previously used to assess the buffering ability of water. However, use 50 mL of the assigned solution instead of distilled water. Record your results in the data table provided.

mL of HCl Added to Buffer Solution	pH
0 (initial pH)	
2	
4	
6	
8	
10	
12	

Calculate the percent change in pH with the acid for activities 3 and 4. To do this, subtract the initial pH from the final pH, then divide that by the initial pH value and multiply by 100 to get a percentage as shown below.

$$\text{Percent change} = \frac{\text{initial} - \text{final}}{\text{initial}} \times 100$$

For example, if the initial pH is 8.0 and the final pH is 7.2 the percent change in pH is 10%.

$$8.0 - 7.2 = 0.8 \qquad 0.8 \div 8.0 = 0.1 \times 100 = 10\%$$

1. How well does distilled water buffer the acid? ________________

2. How well did your solution buffer the acid? ________________

3. Which solution was the best buffer for the acid? ________________

Laboratory Report Sheet

NAME______________________________ **SECTION** ____________ **GRADE** ____________

pH and Buffers

1. What are **two** differences between an acid and a base?

2. What is the difference between ionization and dissociation?

3. What are the differences between a strong acid and a weak acid? Give an example of each.

4. What does it mean when a buffer "can no longer do the job"?

5. Use this equation to answer the following questions regarding the buffering of H ions.

   $CO_2 + H_2O \leftrightarrow H_2CO_3 \leftrightarrow HCO_3^- + H^+$

   a. What will happen to the amount of CO_2 produced when buffering in response to a decrease in pH?

   b. What will happen to the respiratory rate in response to the CO_2 change when buffering a decrease in pH?

c. What will happen to the amount of CO_2 produced when buffering in response to an increase in pH?

d. What will happen to the respiratory rate in response to the CO_2 change when buffering an increase in pH?

6. In the buffering experiment you tested various solutions and their ability to buffer acid. Answer the following questions about that experiment.
   a. What was the question or hypothesis?

   b. What was the dependent variable?

   c. What was the independent variable?

7. What ingredient in Alka-Seltzer™ might be responsible for its buffering ability?

3 Biochemistry and Carbon Compounds

Objectives

To be able to

1. Prepare, interpret, and design positive and negative controls in experiments.
2. Properly use pipettes in laboratory exercises.
3. Identify the reagents that detect starch, sugars, and proteins.
4. Design and conduct biochemical tests on unknown foods and analyze the results to determine the classes of biochemical molecules they contain.

Terms to Define

1. Atoms ____________________
2. Positive Controls ____________________
3. Negative Controls ____________________
4. Monomers ____________________
5. Polymers ____________________
6. Lipids ____________________
7. Carbohyrates____________________
8. Proteins ____________________
9. Amino Acids ____________________

Introduction

Chemistry is the branch of science that studies the properties of, and changes in, matter. Matter has mass, occupies space, and is made up of **atoms. Atoms** are the smallest units of matter. Usually matter is made up of molecules—two or more atoms bonded together.

The study of the chemistry of living things (organisms) is called biochemistry. In organisms, we find four classes of biochemical molecules: **lipids, carbohydrates, proteins,** and **nucleic acids.**

You will learn how to detect the first three groups of biochemical macromolecules in unknown foods and solutions during lab today.

Lipids: include fats, oils, waxes, and sterols.

Carbohydrates: include sugars, starch, cellulose, and glycogen.

Proteins: include enzymes, transport proteins, membrane proteins, toxins, and structural proteins (e.g., collagen, keratin, hemoglobin, fibrin). **Amino Acids** are monomers of proteins.

Nucleic Acids: include DNA, RNA, coenzymes (NADPH, NADH, FADH), and ATP.

Most of these examples are large molecules (**polymers**) that are made up of repeating smaller molecules (**monomers**). The tests you perform today will include indicators that will identify the presence of lipids, sugars, starches, or proteins. To identify the effect of each indicator, you will perform a **positive control** and a **negative control**. Look at Table 3.1. Review the positive and negative reactions of each biochemical molecule. Examine the **indicator** (tests used to determine the presence of a biochemical molecule) we will use in today's lab.

Positive Controls and Negative Controls

A **positive control** uses a sample that contains the substance you're testing for and *gives a positive result.* A **negative control** uses a sample that you know does not contain the substance you're testing for (in some cases, plain water) and *gives a negative result.* The unknown sample is tested and the results compared to the positive and negative control tests to determine the presence or absence of the tested substance.

Table 3.1 contains the positive and negative reactions for the biochemical molecules and indicators you will use in today's lab.

Table 3.1 **Biochemical Molecule Indicators and Positive and Negative Reactions**

Biochemical Molecules	Indicators	Special Instructions	Positive Reactions	Negative Reactions
Lipids	brown paper	use a small drop of vegetable oil	clear paper	opaque paper
Sugars	Benedicts Reagent	boil for 2 minutes	yellow, orange	blue color
Starches	Lugol's Iodine	don't get it on your skin or clothes	black, purple	yellow, amber
Proteins	$CuSO_4$ & 10% NaOH*	$CuSO_4$ is Toxic NaOH is Caustic	purple color	blue color

*** NaOH is Sodium hydroxide. Extra care should be taken when using NaOH because it is a very strong base and can cause severe chemical burns on your skin or eyes.**

Predictions

In today's lab you will be running tests on the food samples you brought to class as well as the positive and negative controls provided. Before you get started, go to Table 3.2 and predict the presence or absence of macromolecules for each food (under the column labeled "Predictions").

Proper Use of Pipettes

For many chemical tests, pipettes are used to measure and dispense solutions. There are several varieties of pipettes available. We will use a disposable pipettes in this lab. This is how to properly use a disposable pipette:

1. Squeeze the pipette bulb before lowering the pipette into the solution.
2. Lower the tip of the pipette into the solution and slowly release the pressure on the bulb. Watch the level of the solution rise in the pipette until it reaches your desired volume.
3. Raise the pipette out of the solution and discharge it by squeezing the bulb.
4. Label the pipette (with a marking pen) to indicate the solution it corresponds to. Use one pipette for each solution. Do not cross-contaminate solutions by using the same pipette in two or more solutions or samples!

ACTIVITY 1: Lipids: Fats and Oils that Provide Long-Term Energy Storage for Organisms

Materials

brown paper squares
distilled water bottle
dropper bottle of vegetable oil
toothpicks
blow dryers

Procedure

1. Take a square of brown paper and use a pencil to draw a line down the middle. On the bottom right side of the paper, write the word *oil*. On the bottom left corner write the word *water*.

2. Place one very small drop of oil on the paper above the word *oil*, and place one drop of water on the paper above the word *water*. Spread each of the drops with opposite ends of a toothpick. You can further speed the drying time by using a small hair dryer.

3. When dry, hold the square up to the light and look through the paper. You can see through the paper with the oil, the oil filled the spaces between the paper fibers (**positive control**). The side with the water is dry and you cannot see through it (**negative control**).

4. Take another square of brown paper and conduct lipid tests on the four food samples provided plus any foods your group brought to test. Fill in Table 3.2 with the names of your samples.

5. Conduct the lipid test on each of the food samples. Record your test results for each food in Table 3.2 under the "Results" column.

ACTIVITY 2: Sugars: Five or Six Carbon Compounds that Provide Short Term Energy for Cells

Materials

hot plate	400-ml beaker	boiling stones
4 test tubes	test tube rack	test tube holder
distilled water bottle	Benedict's reagent	marker pen
5% glucose solution	3-ml disposable pipettes	

Procedure

1. Fill the beaker half full of water and place it on the hotplate (make sure there are a few boiling stones in the beaker). Turn the hot plate heat control knob to #7. In a few minutes the water will boil (this is your boiling water bath).
2. Pick up two test tubes. Near the top of one test tube, use the marker provided to write +. Near the top of a second test tube write –. In the first test tube, pipette 2.0 ml of glucose solution. In the second test tube, pipette 2.0 ml of distilled water.
3. Pipette 2.0 ml of Benedict's reagent into each test tube. Gently stir the test tubes by rolling the tube between your hands.
4. Record the color of each test tube below:

   Positive control (+) test tube: ______________ Negative control (–) test tube: ______________
5. Place the two test tubes in the boiling water bath for 2 minutes, then use a test tube holder to remove the test tubes and place them in a test tube rack on your bench. Record the colors of each test tube: Positive control (+) test tube: ____________________ Negative control (–) test tube: ________________

Questions

1. What solution was in the positive control?______________

   What color was it after it was heated?______________
2. What solution was in the negative control?______________

   What color was it after it was heated?______________

Procedure

1. Pick up a new test tube for each of the four foods provided plus any foods your group brought to test. Label them with the names of your test foods. Add 2 ml of each food sample to the corresponding test tube. Now conduct a sugar test on these samples by adding 2 ml of Benedict's reagent, stirring, and boiling for 2 minutes. Examine Table 3.1 to interpret your test results. Record your biochemical test results for sugars in Table 3.2.
2. Turn your hot plate off when you have finished this experiment!

ACTIVITY 3: Starches

Materials

4 test tubes
pipettes
Lugol's Iodine
starch solution
distilled water
test tube rack

Procedure

1. Pick up two new test tubes. Label one test tube *starch* and a second test tube *water*.
2. Shake the starch solution in the container and pipette 2.0 ml of starch solution into the first test tube (positive control).
3. Pipette 2 ml of water into the second test tube (negative control).
4. Carefully add 3 drops of Lugol's Iodine to each test tube (Caution: Don't spill Iodine, it will stain your clothes and skin).
5. Record the resulting color of each test tube: Starch test tube: ____________ Water test tube: ____________
6. Pick up four new test tubes and conduct a starch test (by adding 3 drops of Lugol's Iodine) to four food samples plus any foods your group brought to test. Use 2 ml of each food in a test tube. Record your test results for starch in Table 3.2.

Questions

1. What did the positive control for starch contain? ____________

   What color is the test tube? ____________

2. What did the negative control for starch contain? ____________

   What color is the test tube? ____________

ACTIVITY 4: Proteins

Materials

4 test tubes
0.5% CuSO4 (copper sulfate) solution
10% NaOH solution (Caution: this can cause very serious chemical burns)
test tube rack
egg albumin solution (a protein)
distilled water bottle
safety goggles (a must when performing the protein test)

Procedure

1. Pick up two test tubes. Label one test tube *egg* and another test tube *water*.
2. Pipette 2.0 ml of egg albumin (a **protein** source) into the first test tube and 2.0 ml of distilled water into the second test tube.

3. Slowly and carefully add 20 drops of NaOH to each test tube. **Use protective goggles here and caution.** NaOH can cause severe chemical burns.
4. Add 4 drops of copper sulfate ($CuSO_4$) to each test tube and gently stir the test tubes by rolling them between your hands. In a few minutes, you will notice a color change in the first test tube.
5. Record the colors of each test tube: Egg albumin test tube: ______________________

   Water test tube: ______________________
6. Which test tube is the positive control? ____________ What color is it? ____________
7. Conduct the protein tests on the four food samples provided and any other foods your group brought in to test. Record your biochemical test results for protein in Table 3.2.
8. Discard the liquid contents of the tubes from all of your tests in the waste containers provided. Dispose of all your test tubes (with any remaining solid material) in the waste bags provided.
9. Collect the data from other classmates to complete Table 3.2.

Questions

1. Which test tube is the positive control? ____________

   What color is it? ____________
2. Which test tube is the negative control? ____________

   What color is it? ____________

Table 3.2 Biochemical Test Results

Sample Food	Predictions				Experimental Results			
	Lipid	Sugar	Starch	Protein	Lipid	Sugar	Starch	Protein
1. Chicken								
2. Tofu								
3. Potato								
4. Fish								
5.								
6.								
7.								
8.								

Analysis

1. Which biochemical test results agreed with your predictions? ____________________

2. What did you learn from doing these tests? ____________________

   ____

3. All lab procedures have their positive features and their limitations. Some are tedious and long, others are quick and easy; some are delicate and expensive; yet others inexpensive. What are two positive features of the lab procedures you did today? ____________________

   ____

4. What are two limitations of the lab procedures you did today? ____________________

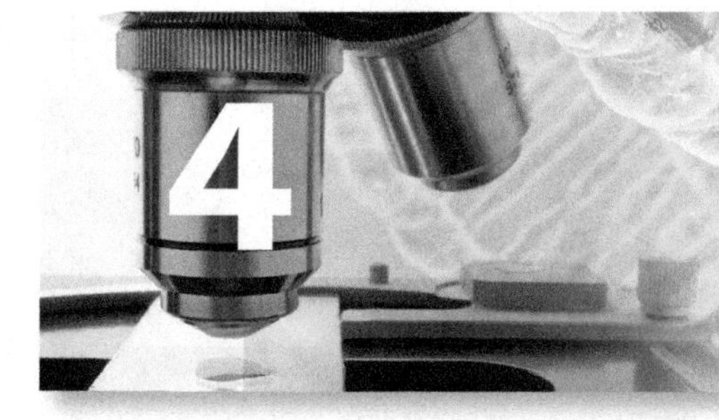

4 SCOPE AND CELLS

Biological molecules come together in complex ways to form cells—the fundamental unit of life. The sizes, shapes, and functions of cells are myriad. A good way to take a quick survey of cells in the various kingdoms is to use the compound light microscope.

Objectives

To be able to

1. Use the compound light microscope to observe the sizes and shapes of cells of bacteria, protists, plants, fungi, and animals.
2. Identify or list:
   a. which kingdoms are prokaryotic vs. eukaryotic.
   b. which features are common to all cells.
   c. internal components of eukaryotic cells.

The Microscope

The compound light microscope (aka scope) is a mechanical device. It does not have any digital components. You must manipulate this machine, using the appropriate parts, to make it work for you. Use the scope carefully but don't be afraid to use the moving parts.

How Tos

How to Carry the Scope

1. Always use both hands.
2. Carry with one hand under the base and the other in the recessed handle if available.
3. Never slide the scope across a surface; pick it up to reposition it.

How to Set up the Scope

1. Set the scope gently on bench top.
2. Remove the dust cover; fold and put it out of the way.
3. Unwrap the cord from body of scope; plug into outlet.
4. Stage should be positioned at the midpoint of its range.
5. Lowest (shortest, red-lined, 4×) objective lens should be in position.
   a. If not, use the revolving nosepiece to position it; it will click into place.

How to Get Started

1. Turn on the light. Note that the light comes from beneath the stage.
2. Turn the rheostat (dimmer switch); observe the change in light intensity.

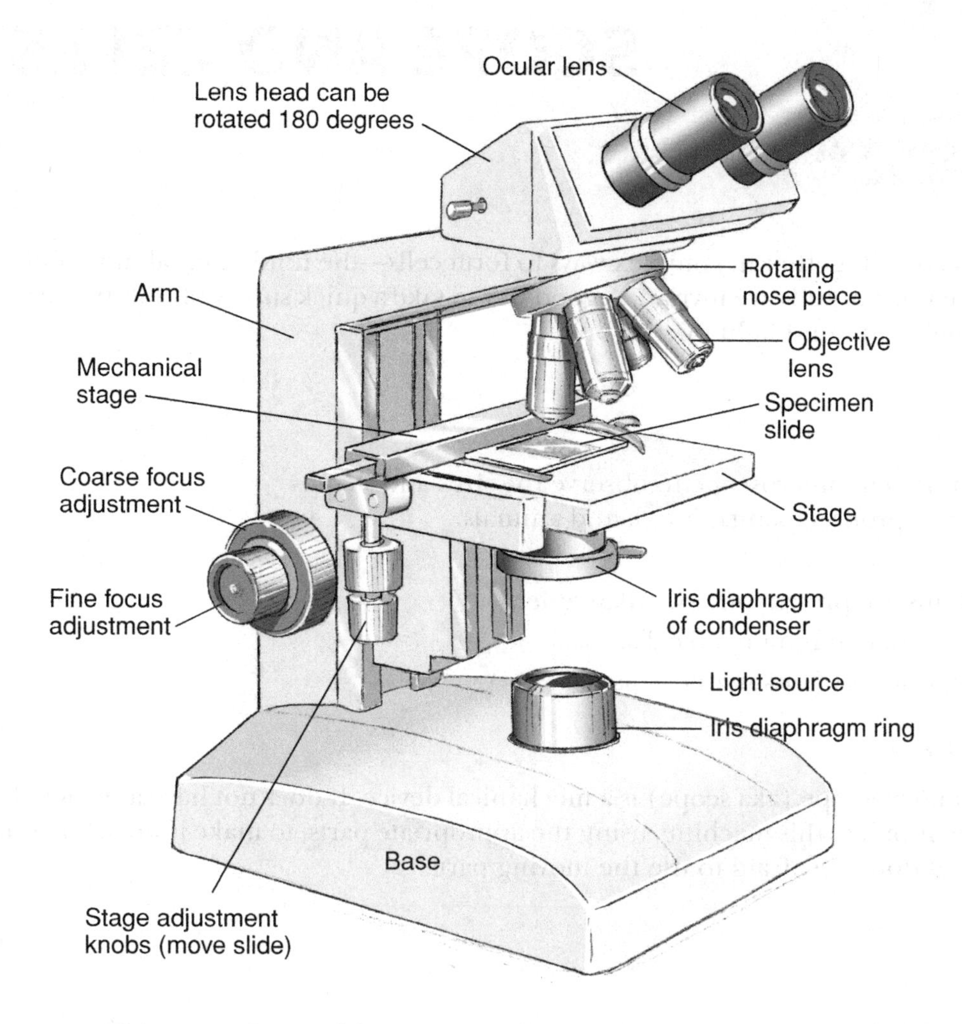

Figure 4.1 Parts of the compound microscope.

CARE AND SAFETY

1. Use only lens paper to clean the lenses. Never use paper towel or tissue. Ask your instructor, who has lens paper and cleaning fluid, to assist you.
2. Always use a coverslip. DO NOT allow water and the salt or dye it contains to touch the lens or the stage. You may carefully blot a small spill on the stage dry, using a small piece of paper towel. If liquid is near the objective lens, ask your instructor for assistance *immediately*. Otherwise the lenses may be permanently damaged.
3. Minor scratches or even cracks in the coverslip will not impede your work; you will focus past them.

How to Adjust the Scope for Your Face (Interpupillary Distance)

1. Look through the ocular lenses. (No specimen necessary.)
2. Use the rheostat to adjust the light intensity for your comfort.
3. Hold the ocular lenses with both hands and adjust their position (as you would a pair of binoculars) until you see one circle of light (see figure below, left).
   a. This may not be entirely comfortable, but will become more so with practice.
   b. If you look through just one ocular, you will experience eyestrain.
4. If the field of view (the circle you see) is only partially filled with light, slowly adjust the field iris diaphragm by moving the ring until the field of view is filled (see figure below, right).
   a. At higher magnifications you will readjust the field iris diaphragm.

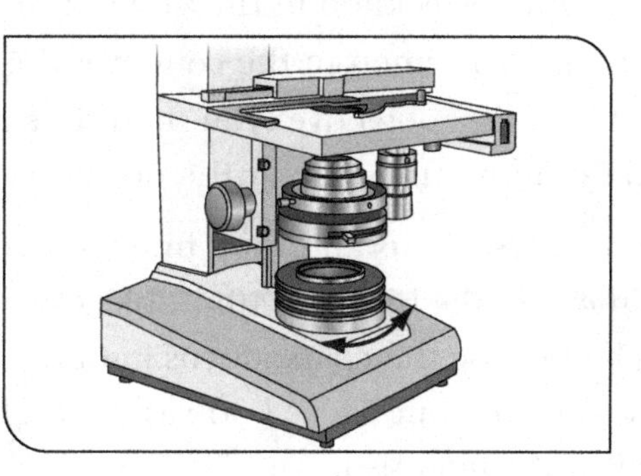

Figure 4.2 How to adjust the scope for your face.

How to Put a Slide on the Stage (see Figure 4.3)

1. Make sure the stage is low enough to accommodate placing the slide.
2. Make sure the lowest objective is in place.
3. Open the specimen holder by moving the curved finger knob.
4. Place the slide on the stage and slide it back as far as it will go to the lower left.
5. Gently release the curved finger knob.
6. DO NOT put the curved finger on top of the slide.

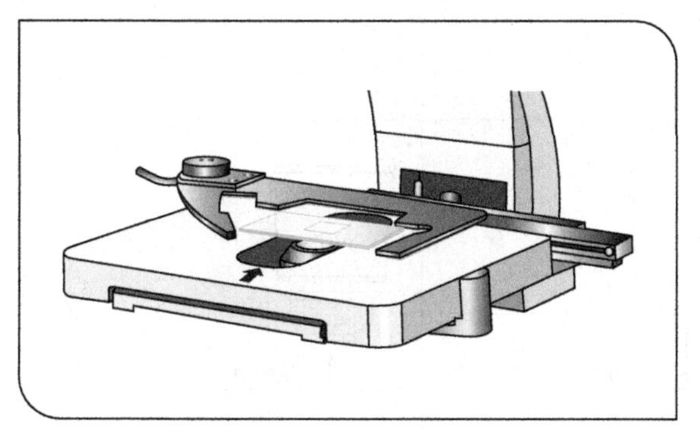

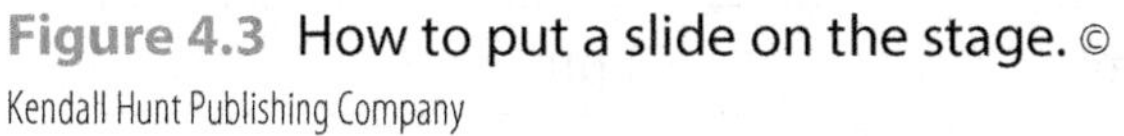

Figure 4.3 How to put a slide on the stage.

Figure 4.4 How to move the specimen.

How to Move the Specimen (see Figure 4.4)

1. Move the mechanical stage adjustment knobs.
2. DO NOT move the stage with your hand.
3. Position the slide so that the specimen is centered over the light. If in doubt, start by centering on the middle of the coverslip.

HOW TO FOCUS ON A SPECIMEN

1. Always start with the lowest (4×) objective engaged.
2. Insert the slide as above.
3. Using the coarse focus knob, position the stage as close as possible to the objective lens.
4. While looking through the oculars, use the coarse focus knob to slowly move the stage toward the objective lens until you can see the specimen.
5. To make sure you have the best possible focus, continue using the coarse focus until you go past the best focus and then back to the best.
6. Use the fine focus knob in the same way to get the best possible view.
7. Position the specimen in the center of the field of view.
8. Move to the 10× objective (you should be able to hear it click into place).
   a. Do not move the stage or the position of the slide on the stage at this point.
9. Use the coarse focus and then fine focus, as before.
10. Next move to the highest (40×) objective.
    a. DO NOT use the coarse focus knob.
    b. Use only the fine focus knob at the highest objective.
    c. Adjust the light using:
       i. aperture iris diaphragm by moving the lever.
       ii. field iris diaphragm by moving the ring.

Table 4.1 Comparison Among Objectives.

Objective	4×	10×	40×
Total magnification	40×		
Diameter of field of view	5 mm (=5000 μm)	2 mm (=2000 μm)	
Working distance		10x 8.3 mm	40x 0.5 mm
Depth of view	175 μm	28 μm	3 μm

Important Concepts

Magnification Is Multiplicative

When you are using the 4× objective lens, the image is magnified four times. Then the ocular lens magnifies that image 10 times, for a total of 40×. Fill in the total magnification for each objective lens in Table 4.1.

The Diameter of the Field of View Decreases with Increasing Magnification

At 40×, the diameter of the field of view is 5 mm (see Table 4.1). At 100×, the diameter of the field of view is 2 mm. At 400×, which is a tenfold greater magnification than at 40×, the diameter of the field of view is ______________________________

____

The Working Distance Decreases with Increasing Magnification

In other words, there is more space between the objective lens and the specimen at 4× than 10×, and less space the higher the objective lens. At 400×, the space is only 0.5 mm (see Table 4.1). Why must we **not** use the coarse focus knob at total magnification of 400×? ______________________________

____

The Depth of View Decreases with Increasing Magnification

The distance through which you can focus is 175 μm at 40× total magnification, but only 3 μm at 400× total magnification. If you want to focus through an object, which magnification should you use?

____

The Objectives Are Parfocal

When you move to the next highest objective, the specimen should already be focused enough to be discernable. You should have to do minimal focusing to get the image in focus with the new objective. Once you have focused well at 40×, should you move the stage down before going to 100×? ______________________________ What about going from 100× to 400×?

____

The Objectives Are Parcentric

This is why you should center the specimen in the center of the field of view. When you move to the next highest objective, it is the center of the field of view that is further magnified. If the specimen of interest is at the edge of the field of view, it will not be visible when you increase magnification. If you "lose" the specimen, which would be the best approach: start over at the lower magnification or search at the higher magnification?

____

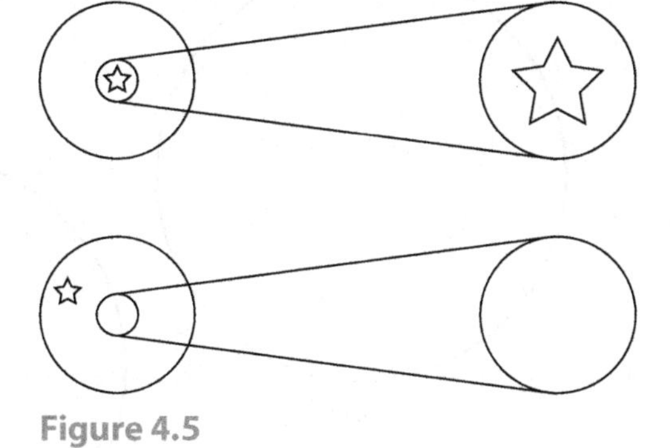

Figure 4.5

Practice Using the Scope

Two prepared slides, "Letter e" and "Colored Threads," can both be used to practice focusing on an object. Some points described above can be illustrated with either slide. Different additional points can be made with "Letter e" and "Colored Threads."

Because the diameter of the field of view decreases as the magnification increases, the amount of the specimen that you can see also decreases.

At higher magnification, as the working distance decreases, the image contrast will also decrease. You can improve the image by adjusting the light, using the iris diaphragm lever or the field diaphragm ring. Be aware that contrast (perceivable difference between two objects) and resolution (ability to discern detail) cannot be simultaneously maximized. Use the light and fine focus to obtain the best image for your purposes.

Colored Threads Slide

1. Hold the slide up to the light. Notice that three threads (blue, red, and yellow) are visible with the naked eye.
2. Using the "How to focus on a specimen" procedure, focus on the specimen at 40× total magnification.
   a. Remember to use the stage adjustment knobs to center the specimen.
   b. Can you bring all three threads into focus at the same time? __________
   c. Focus through the threads using the fine focus.
      i. Can you tell which thread *appears* to be on top? __________
   d. Sketch what you see to scale in the space below. "To scale" means draw it as you see it. If the specimen takes up a third of the diameter of the field of view, draw it that way.
3. Focus at 100× total magnification:
   a. Can you see all three threads? __________
   b. Can you see more or less of the specimen than at 40×? __________
   c. Sketch what you see to scale.
4. Focus at 400× total magnification:
   a. Can you see all three threads? __________
   b. Can you bring all three threads into focus at the same time? __________
   c. Can you focus through as much as you could at 100× or 40×? __________
   d. Sketch what you see to scale.

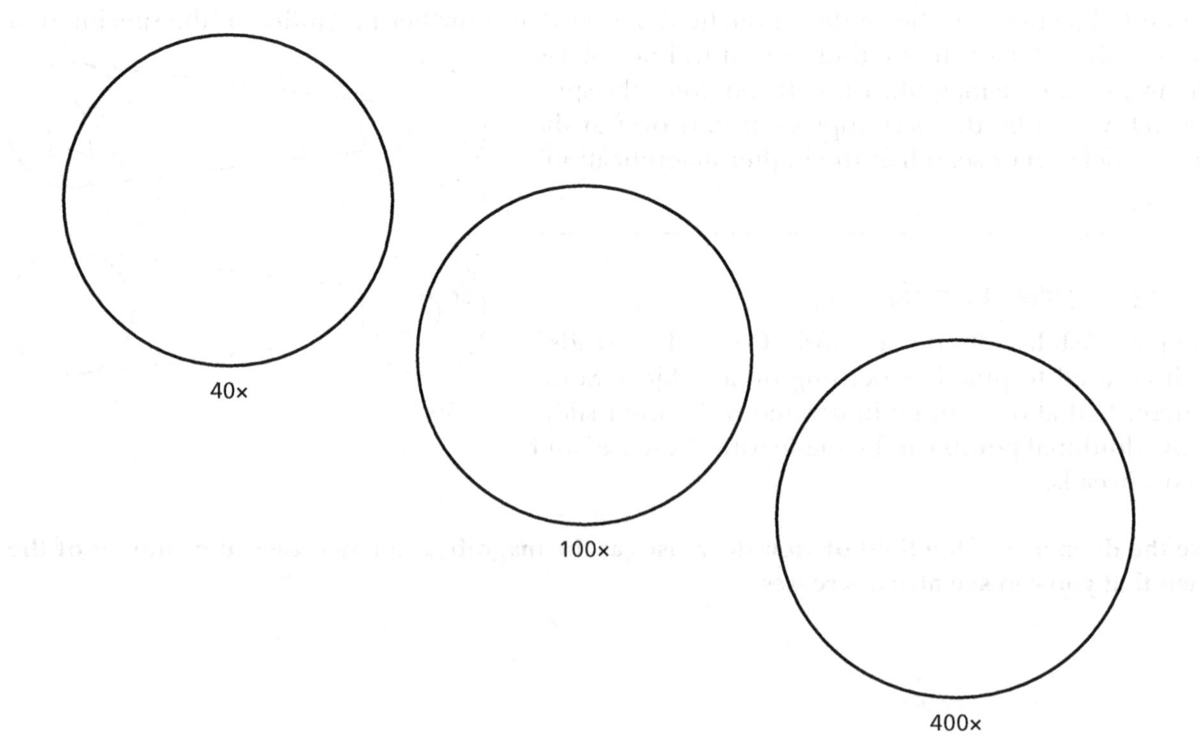

Letter e Slide

1. Hold the slide up to the light; the specimen is visible with the naked eye.

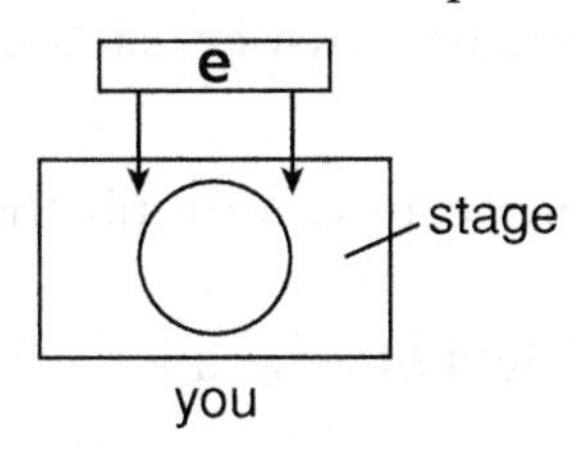

Figure 4.6

2. Orient the slide so that the letter 'e' is right-side up. Place the slide on the stage this way.
3. Using the correct procedure, focus on the specimen at 40× total magnification.
   a. The mirrors in the scope flip and reverse the image before it reaches your eyes.
   b. Compared to the right-side up 'e' you saw with the naked eye, the magnified image is ______ ______________________________.
   c. Draw it to scale in the space below.
4. *When you move the stage in one direction, the image moves in the other direction.*
   a. But you won't really need to think about this as you proceed; your brain will adapt as you use the scope.
5. Focus at 100×.
   a. Sketch what you see to scale in the space below.
6. Focus at 400× total magnification.
   a. Can you see the entire letter? __________
   b. Sketch what you see to scale.

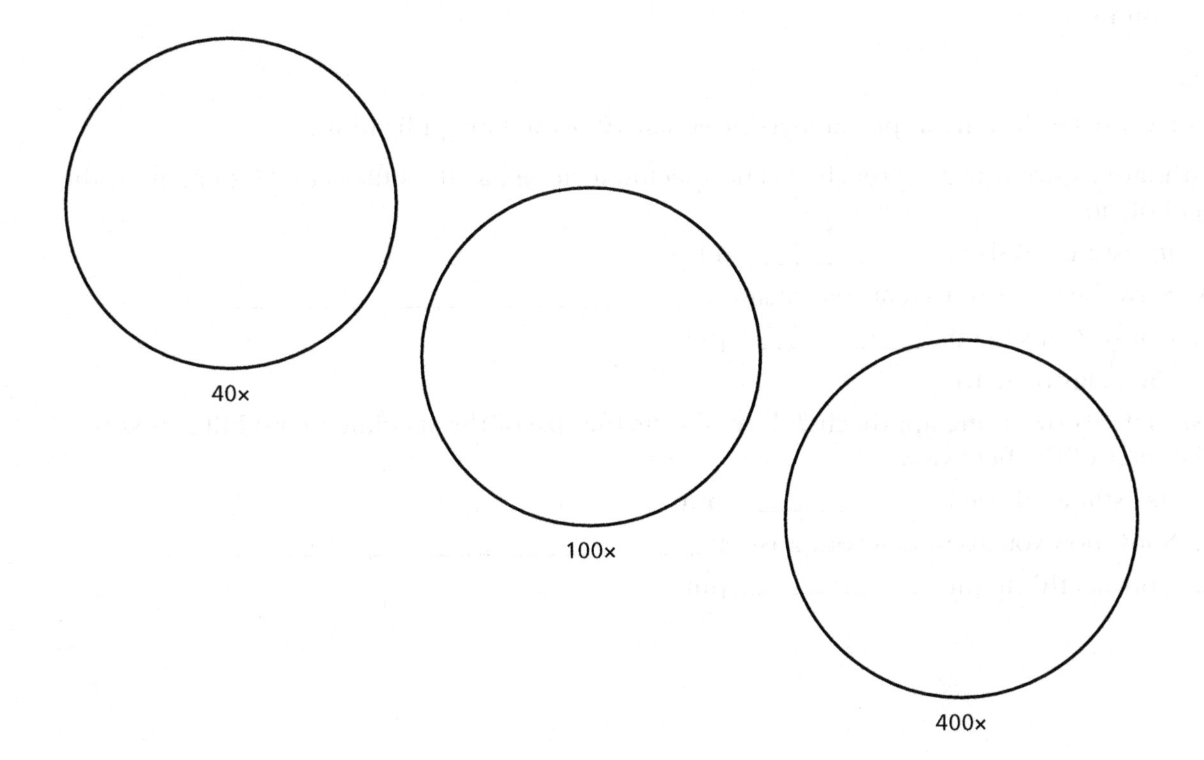

Estimating the Size of a Specimen

Since we know the diameter of the field of view (see Table 4.1), it is easy to *estimate* the size of a specimen. There are two approaches, *which are equivalent.* You should get the same answer using both approaches.

1. What proportion of the diameter of the field of view does the specimen occupy? Multiply that proportion by the diameter of the field of view. See Figure 4.7.
   a. For example, at 40×, if specimen takes up half the diameter of the field of view, it is [(0.5)*(5 mm)] or 2.5 mm in size.

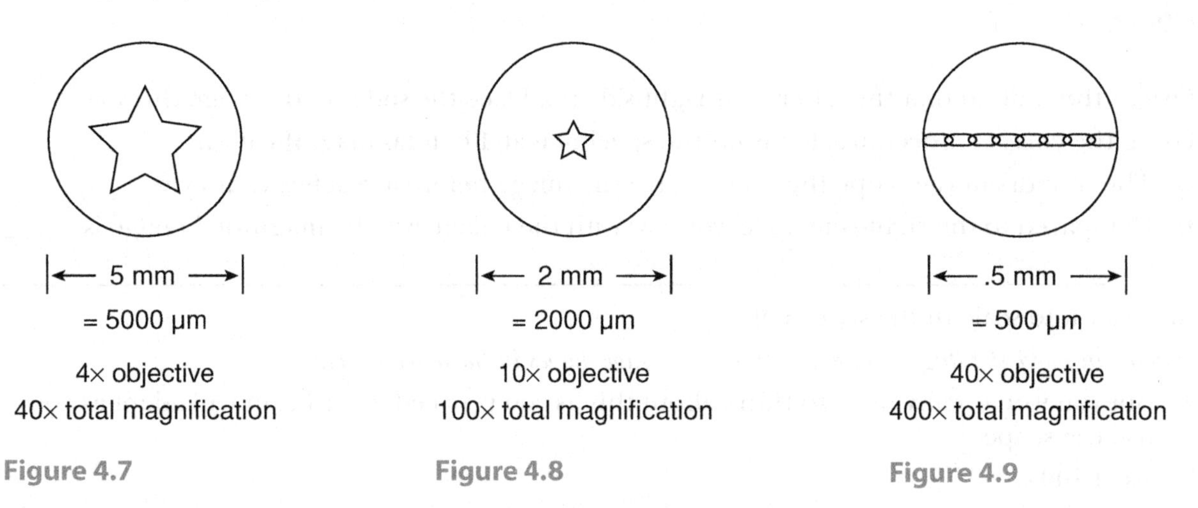

Figure 4.7 Figure 4.8 Figure 4.9

2. How many of the specimens could fit across the diameter of the field of view? Divide the diameter of the field of view by that number. (Still referring to Figure 4.7.)
   a. For example, at 40×, if the specimen occupies half of the field of view, then two objects the size of the specimen could fit across the diameter of the field of view. It is [(5 mm)/2] = 2.5 mm in size.

Practice

Look at Figure 4.8 showing a specimen as viewed at 100× total magnification.

1. Estimate its size using approach 1: The specimen viewed at 100× takes up 20 percent of the field of view.
   a. Its estimated size is ____________ mm.
   b. Show how you arrived at your answer. ______________________________
   c. Convert this to μm. ____________ μm
   (See Figure 4.11.)
2. Estimate its size using approach 2: Five objects the size of the specimen could fit across the diameter of the field view.
   a. Its estimated size is ____________ mm.
   b. Show how you arrived at your answer. ______________________________
   c. Convert this to μm. ____________ μm

Look at Figure 4.9 showing a specimen as viewed at 400× total magnification.

3. At 400×, eight objects the size of the specimen could fit across the diameter of the field of view. In other words, the specimen takes up about 12% of the field of view. Use either method to estimate the size of the specimen.
   a. The estimated size is ____________ mm.
   b. Show how you arrived at your answer. ______________________________
   c. Convert this to μm. ____________ μm

You must use the appropriate magnification to make your estimate. If the specimen takes up most of the field of view, use the next lower magnification. If the specimen is so small that it is difficult to estimate how many would fit, or what proportion of the field of view it occupies, use the next higher magnification. You may not be able to reasonably estimate the size of a specimen. (This is the case for the Eubacteria and Archaea cells which are 1–10 μm.)

Cells

Prokaryotes

The term *prokaryote* means "before kernel" and refers to cells that do not have a nucleus. All prokaryotic cells are single-celled, and in addition to lacking a nucleus, they also lack much else in the way of internal organization. The term *prokaryote* generally encompasses two kingdoms, the Eubacteria and the Archaea. We will look at examples only of Eubacteria.

Observe the three demonstration microscopes.

1. Prepared slides of bacteria are already set up at 400×.
2. You will need to adjust the fine focus in order to discern the shapes.
3. DO NOT move the coarse focus, the objective lens, or the stage.
4. The cells are small enough (1–10 μm) that you will not be able to accurately estimate the size or make out details (or lack thereof, i.e., little or no internal organization).
5. Complete the table below.

Table 4.2 The Three Major Bacterial Shapes

Shape	Coccus	Bacillus	Spirillum
Written description	spherical	rod-shaped	spiral
Drawing (not to scale)			

Eukaryotes

Eukaryotic organisms may be single-celled or multi-celled. They have, in addition to the nucleus, several other membrane-bounded organelles, which are specialized structures that perform specialized functions. Examples include mitochondria and chloroplasts. You will easily see chloroplasts in the aquatic plant *Elodea.* But mitochondria are small enough that, even when dyed, it takes time and effort to observe them with the light microscope. Eukaryotic cells also have an extensive cytoskeleton made of linear protein elements. You will not be able to directly observe the cytoskeleton today, but will see evidence of it if you observe cytoplasmic streaming—the flowing of cytoplasm that enables the movement of molecules within the cell. Eukaryotes include the kingdoms Fungi, Plants, Animals, and Protists. To look at an example of each of these groups, you will make wet mounts.

HOW TO MAKE A WET MOUNT

1. Place a drop of suspension (if fungus or protist) or a drop of water and then the specimen (if plant or animal) on the slide.
2. Place the coverslip as shown below. Slowly lowering the coverslip will take advantage of the adhesive and cohesive properties of water to minimize air bubbles.
   a. Air bubbles large enough to be seen with the naked eye may impede your ability to view your specimen. Gently tap on the coverslip to remove them, or remove the coverslip and try again.
   b. Small air bubbles, which may appear as very regular shapes with a dark edge, may be annoying but you should be able to work around them.

   Before you place the slide on the stage, make sure the bottom of the slide is dry.
3. If, when you view your wet mount, it appears to be vibrating, remove the slide and use a small piece of paper towel to wick up some of the excess water.
4. When you are finished with the wet mount, rinse and dry the slide and coverslip.

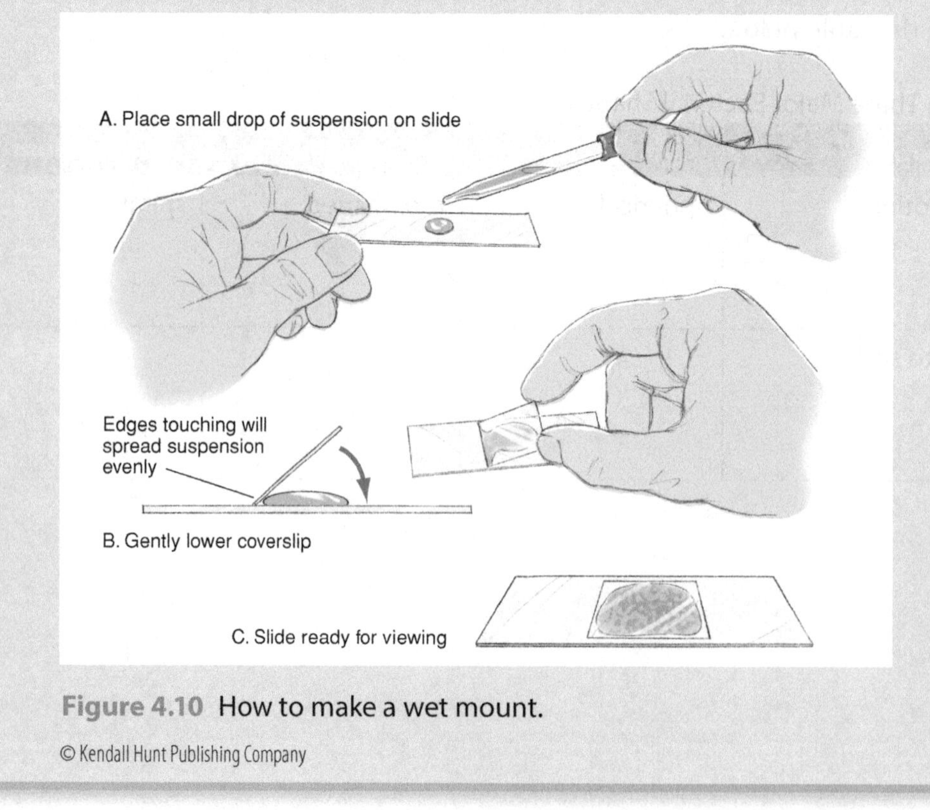
A. Place small drop of suspension on slide
Edges touching will spread suspension evenly
B. Gently lower coverslip
C. Slide ready for viewing

Figure 4.10 How to make a wet mount.

Plant, Animal, Fungus, Protist

For each of the live samples, *use the most appropriate magnification* to

1. Make a drawing (to scale; draw what you see).
2. Estimate the size of a single cell.
3. Give a written description of the shape of the cells.

Plant

1. Make a wet mount of the smallest developing leaves at the growing tip (apical meristem) of the freshwater aquatic plant *Elodea*.
2. In your drawing, label the cell wall and chloroplasts.

Plant
Total magnification ______________ Size estimate ______________

Description ____

____

Animal

1. Make a wet mount of your cheek epidermal cells.
   a. Sterilize a toothpick by wiping it with an alcohol swab.
   b. Use the toothpick to gently scrape the inside of your cheek to remove cells that have already sloughed off or are ready to slough off. DO NOT break or tear the skin.
   c. After you have transferred the cells from the toothpick to the slide, dispose of the toothpick in the disposal container provided. DO NOT lay your used toothpick on the bench.
2. Stain the specimens by adding a small drop of methylene blue. This will enable you to more easily see the cells and the nucleus in each cell.
3. In your drawing label the nucleus.

Animal
Total magnification ____________ Size estimate ____________
Description ______________________________________
____

Fungus

1. Make a wet-mount of the suspension containing *Saccharomyces cerevisiae*, commonly called baker's yeast. (Many fungi, including mushrooms, are macroscopic.) You can add a drop of water to your slide to thin the suspension.

Fungus
Total magnification ______________ Size estimate ______________

Description ____

____

40×

100×

400×

Protist

1. Make a wet mount of the pond water and scan to find protists.
   a. These may include small clusters of single-celled algae, diatoms, colonial photosynthetic *Volvox*, protozoans such as *Amobea*, and ciliated *Paramecium* and *Vorticella*. You may also see larger organisms such as rotifers and nematodes, which are microscopic animals.
2. Include as much detail as possible in your drawing.

Protist
Total magnification ______________ Size estimate ______________

Description ____

____

CARE AND SAFETY

When You Are Done with the Microscope

1. Rinse and dry slides and coverslips; return to box.
2. Put the scope away properly.
   a. Make sure there are no slides on the stage.
   b. Position the stage at its midpoint.
   c. Use the revolving nosepiece to engage the 4× objective.
   d. Wrap the electrical cord securely around the body of the scope.
   e. Put on the dust cover.
   f. Place carefully on storage shelf.

Cell Components

The light microscope enables you to see both prokaryotic and eukaryotic cells, and to see some structures within eukaryotic cells. But there are many structures that cannot be easily visualized. It is important to have a general knowledge of cell structures and their functions. Use Table 4.3 to answer the questions below.

Which components are common to all cells?

___

___

___

Which kingdoms are eukaryotic?

___

___

___

___

Which components are found in plants but not animals?

___

___

___

Table 4.3 Cell Components

	Prokaryote	Eukaryote			
		Protists	Fungi	Plants	Animals
Cell wall	√	√ (many)	√	√	
Cell or plasma membrane	√	√	√	√	√
Ribosomes	√	√	√	√	√
DNA	√	√	√	√	√
Membrane-bounded organelles			√		
Nucleus		√	√	√	√
Endomembrane system (rough smooth ER, vesicles, Golgi complex)		√	√	√	√
Mitochondria		√	√	√	√
Chloroplasts		√ (some)		√	
Central vacuole				√	
Cytoskeleton		√	√	√	√

The Size of Things

1 cm = 10^{-2} (1/100) m
1 mm = 10^{-3} (1/1000) m
1 μm = 10^{-6} (1/1,000,000) m
1 nm = 10^{-9} (1/1,000,000,000) m
100 cm per meter
10 mm per cm
1000 μm per mm

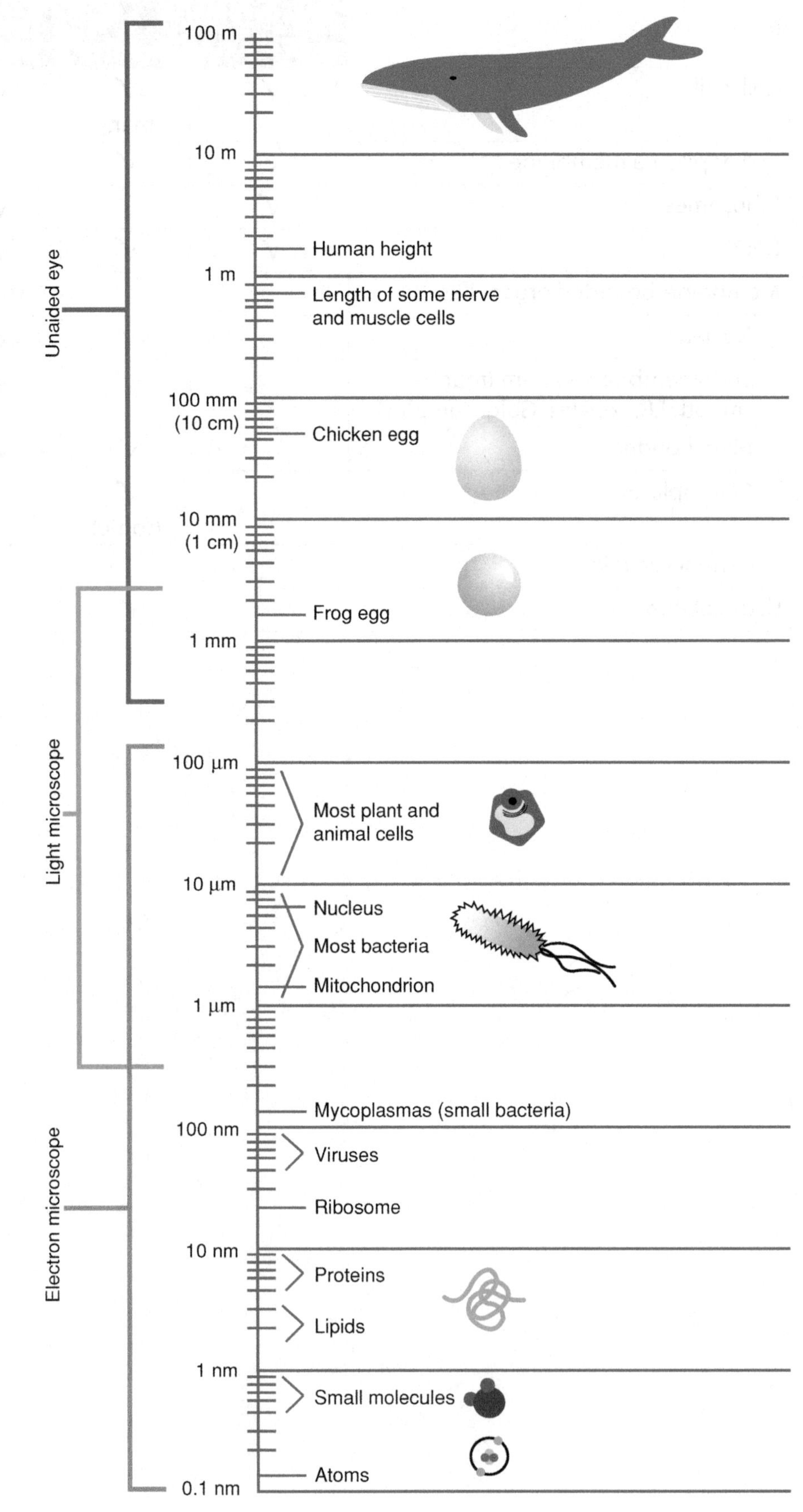

Figure 4.11

5 Enzyme Activity

Objectives

To be able to

1. Identify the conditions most favorable for enzyme activity (e.g., neutralpH)
2. Identify the ingredients and the results of the three enzyme procedures
3. Identify the type of organic molecules enzymes are grouped with

Terms to Define

1. acid __________
2. base __________
3. pH __________
4. neutral pH __________
5. enzyme __________
6. catalase __________
7. urease __________
8. rennin __________
9. metabolism __________
10. metabolic pathway __________
11. substrate __________
12. product __________
13. synthetic __________
14. degradive __________
15. catalyst __________

Introduction

ll the biochemical reactions that occur within an organism are collectively called **metabolism**. These enzyme-aided reactions are organized into **metabolic pathways**. In each metabolic pathway, an enzyme converts **substrate** molecules into a final **product** molecule. Substrate molecules are the actual participant molecules that will either be bonded together in a **synthetic** pathway or broken apart in a **degradive** pathway. An example of a synthetic pathway is the conversion of the common 6-carbon sugar glucose into the 5-carbon sugar ribose, a necessary component of nucleotide molecules (contained in both DNA and RNA).

There are more than 5,000 different known enzymes. Enzymes are protein molecules that act as **catalysts**. A catalyst increases the rate of a chemical reaction that would normally occur at a slower rate. The enzyme molecule carbonic anhydrase, which combines carbon dioxide (CO_2) and water to form carbonic acid (aweak acid responsible for carrying CO_2 in the blood) performs 600,000 reactions persecond. Enzymesmaybeused over and over again without being used up or altered. A particular enzyme, however, is very specific; that is, it can act upon only one or a very few, similar kinds of substrates. This specificity has been compared to the "lock-and-key" fit of complementary base pairs in the DNA molecule.

Enzyme concentration, pH, temperature, and **time** are among the principal factors affecting enzyme activity. In this lab you will experiment with these factors to determine the influence they have.

ACTIVITY 1: Enzyme Concentration

In this experiment you will work with the enzyme **urease**. Urease converts urea, a waste product of complex organisms like mammals, into ammonia and carbon dioxide. The urea solution that you will be working with is amber in color and somewhat acidic. As a result of urease enzyme activity, urea will be converted into carbon dioxide and ammonia, turning the urea solution to a basic pH and a red/pink color. The degree and rate of red/pink color development in the urea solution is a measure of urease enzyme activity.

Procedure for Observing the Effect of Enzyme Concentration on Enzyme Activity

1. Measure 4 milliliters (mls.) of urea solution in a graduated cylinder and pour it into a clean test tube. Place the tube in a test-tube rack. Do not add any urease powder. Mark the tube with a C or #1. This is your control.
2. Measure 4 mls. of urea solution and pour it into a tube marked #2.
3. Measure 4 mls. of urea solution and pour it into a tube marked #3.
4. Add 1 scoop of urease enzyme powder to tube #2.
5. Add 2 scoops of urease enzyme powder to tube #3.
6. Place all tubes in a warm-water bath.

NOTE: If you are using a raw bean mash enzyme source, record the color changes that may occur in the tubes at 10-minute intervals over a 50-minute period. If you are using a purified enzyme, the entire reaction and the color changes will take only 5 minutes, and you should record color development every minute. Record your observations in the table below. If you are using raw bean mash, go on to Procedure 2 while you are waiting for color development.

	0 min.	10 min.	20 min.	30 min.	40 min.	50 min.
Tube #1						
Tube #2						
Tube #3						

Questions

1. At the end of 50 or 5 minutes (depending on your enzyme source), did tubes 2 and 3 develop the same degree of color (intensity) change? ____________________
2. Which tube developed color faster? ____________________
3. Why? ____________________
4. What was the purpose of the control tube?
____

ACTIVITY 2: How Temperature effects Enzyme Activity

Most enzymes, especially those found in the bodies of warm-blooded organisms, are most active at warm temperatures. Cold temperatures generally slow biochemical reactions and hot temperatures will destroy or **denature** the molecular structure of the enzyme. In this part of the lab you will determine the activity of the enzyme **rennin.** Rennin is found in the stomachs of cows and newborn babies. It helps solidify milk so that it can be digested.

Procedure

1. Withdraw 2 milliliters (mls.) of cold milk with a pipette and dispense into a clean test tube. Mark the tube "Cold" with a marker. Add one drop of cold rennin to the "Cold" tube and place this tube in a cold-water bath.
2. Withdraw 2 mls of warmed milk and dispense into a clean tube marked "Warm" Add one drop of warmed rennin to the "Warm" tube and place this tube in a warm-water bath.
3. Withdraw 2 mls. of warmed milk and dispense into a clean tube marked "Boiled." Add one drop of boiled rennin to the "Boiled" tube and place this tube in a warm-water bath.
4. After 15 minutes examine the tubes and record below whether the milk solidified fully, partially, or not at all.

NOTE: After you have set up this part of the lab, go on to Procedure 3 during the 15-minute waiting period.

	Observations
Cold Tube	
Warm Tube	
Boiled Tube	

Questions

1. Why did the warm milk and boiled rennin tube not solidify, even though it was placed in a warm-water bath?

2. Why did the cold milk and cold rennin tube placed in a cold-water bath not solidify?

3. Why did the warm milk and warm rennin tube solidify when placed in a warm-water bath?

ACTIVITY 3: The Effect of pH on Enzyme Activity

Enzymes facilitate the biochemical reactions of metabolic activity within an organism at a specific pH or in a "preferred" pH range. The pH scale is used as a measure of acidity (pH 0.1 through 6.9) or base (pH 7.1 through 14). A pH of 7 is neutral. In this part of the lab you will try to determine the optimum pH at which the enzyme **catalase** (present in the potato) is most active in breaking down hydrogen peroxide into water and oxygen gas. Hydrogen peroxide (H_2O_2) is a toxic metabolic by-product derived from the breakdown of organic molecules in cells. The amount of bubbles evolved in different pH environments will determine the degree of catalase enzyme activity.

Procedure for Determining the Effect of pH on Enzyme Activity

1. Measure 2 milliliters (mls.) of distilled water (having a neutral pH) in a graduated cylinder, and pour it into a clean test tube. Mark the tube "DW" with a marker.
2. Cut a piece of peeled potato into a cube so that each side of the cube measures about 1 centimeter across. Use either your dissecting scalpel or a single-edge razor blade, and measure with a plastic centimeter ruler.
3. After you have cut your potato to these dimensions, crush and mash the piece in a mortar with a pestle. Add this mush to your "DW" tube with the help of a spoonula.
4. Wait three minutes.
5. Measure out 3 mls. of hydrogen peroxide and pour it into the "DW" tube with the mashed potato in it. After five minutes, observe and note the degree of bubbling (some, a lot, or none) in the chart below.

NOTE: Rinse the graduated cylinder three or four times with tap water before measuring a new chemical agent. Rinse spoonula also, between additions of mashed potato.

6. Measure out 2 mls. of a weak hydrochloric acid (HCl) solution, pour it into a clean tube, and mark the tube "Acid". (Be sure to rinse the graduated cylinder with tap water after pouring the HCl).
7. Cut another piece of potato and mash it with a mortar and pestle. Add the mush to the HCl acid tube.
8. Measure out 3 mls. of hydrogen peroxide and pour it into the HCl acid tube. After five minutes, record the degree of bubbling in the chart below.

9. Measure out 2 mls. of a weak sodium hydroxide (NaOH) base solution, pour it into a clean tube, and mark the tube "Base". (Be sure to rinse the graduated cylinder with tap water once again).
10. Cut another piece of potato and mash it with a mortar and pestle. Add the mush to the NaOH "Base" tube.
11. Measure out 3 mls. of hydrogen peroxide and pour it into the NaOH "Base" tube. After 5 minutes, record the degree of bubbling in the chart below.

	Degree of Bubbling
"DW" tube	
HC1 "Acid" tube	
NaOH "Base" tube	

RINSE CLEAN YOUR MORTAR, PESTLE, SPOONULA, AND GRADUATED CYLINDER.

Question

1. Which tube had the most bubbling? ______________________ Why? ______________

____

Laboratory Report Sheet

NAME ______________________ **SECTION** __________ **GRADE** __________

Enzyme Activity

1. In the potato catalase experiment, the hydrogen peroxide acted as the ______________ in the enzyme reaction ______________
   a. catalyst
   b. enzyme-substrate complex
   c. substrate
   d. enzyme

2. Most enzymes are most active in ______________ temperature.
   a. boiling
   b. cold
   c. warm

3. A metabolic degradive pathway involves the ______________ of molecules.
   a. synthesis
   b. breakdown
   c. dehydration
   d. building

4. In the potato catalase experiment, which tube showed the most enzyme activity? ______________
   a. the tube with hydrochloric acid
   b. the tube with distilled water
   c. the tube with sodium hydroxide
   d. the tube with ammonia

5. Rennin is found in the ______________ of newborn babies.
   a. liver
   b. stomach
   c. pancreas
   d. intestine
   e. saliva

6. A ______________ color indicated the activity of the urease enzyme.
   a. pink/red
   b. yellow/brown
   c. clear
   d. yellow

7. Boiling temperatures will destroy or denature the molecular structure of an ______________ enzyme.
   a. True
   b. False

8. A catalyst ______________ the rate of reaction.
   a. increases
   b. decreases
   c. stops
   d. does not effect

9. Urease converts the waste product urea into ______________
   a. water and ammonia
   b. carbonic acid
   c. water and carbon dioxide
   d. carbon dioxide and ammonia
   e. ammonia

10. The milk/rennin experiment was meant to show the effect of ______________ on enzyme activity.
    a. heat
    b. substrate
    c. concentration
    d. pH

11. The enzyme ________________ is present in the potato.
    a. urease
    b. amylase
    c. catalase
    d. rennin
    e. dehydrogenase

12. The term that best describes enzyme specificity is ________________
    a. degradation
    b. catalyst
    c. lock and key

13. The potato catalase experiment was meant to show the effect of ________________ on enzyme activity.
    a. heat
    b. mashing time
    c. substrate
    d. concentration
    e. pH

14. The enzyme catalase is responsible for breaking down the toxic chemical ________________.
    a. arsenic
    b. ammonia
    c. hydrochloric acid
    d. hydrogen peroxide

15. Enzymes are ________________.
    a. proteins
    b. carbohydrates
    c. amino acids
    d. lipids

16. Rennin helps solidify milk so that it can be ________________.
    a. regurgitated
    b. reused
    c. preserved
    d. digested
    e. reserved

17. Ammonia and carbon dioxide are the result of the activity of this enzyme ________________.
    a. catalase
    b. urease
    c. rennin
    d. amylase

18. The term that best describes enzyme activity is ________________
    a. degradation
    b. catalyst
    c. lock and key

19. An enzyme molecule can be used only once ________________
    a. True
    b. False

20. A metabolic synthetic pathway involves the ________________ of molecules.
    a. degradation
    b. breakdown
    c. dehydration
    d. building

21. Enzyme activity is affected by ________________.
    a. pH
    b. temperature
    c. enzyme concentration
    d. all of these
    e. none of these

22. The result of the milk/rennin experiment indicated which combination of circumstances obtained the greatest enzyme activity? ________________
    a. cold milk/cold rennin
    b. warm milk/boiled rennin
    c. cold milk/cold rennin
    d. warm milk/warm rennin

23. In the urease experiment, which tube had the greatest degree of enzyme activity?

    ____

    a. the control tube
    b. the tube with 2 scoops of urease
    c. the tube with 1 scoop of urease
    d. the tube with 3 scoops of urease

24. The pH of distilled water is ________________
    a. acid
    b. base
    c. neutral
    d. alkaline
    e. none of these

6A CELLULAR RESPIRATION

Introduction

In terms of acquiring energy to do the work of the cell, each cell is on its own. Each cell must continuously create its own supply of ATP. To do this, cells convert energy in the form of the biochemical glucose into energy in the form of ATP. In other words, glucose is broken down to create ATP.

Objective

To be able to

1. Explain the process of cellular respiration.

Background

The process by which cells break down glucose is called cellular respiration. It requires nutrient molecules and oxygen. Carbon dioxide and water are the products of the series of reactions involved in cellular respiration.

$$C_6H_{12}O_6 + 6O_2 \rightarrow 6CO_2 + 6H_2O + (36\ ATP)$$

There is another important feature of cellular respiration which is not shown in these equations. Cellular respiration involves many small steps; these multiple steps allow the cell to use the energy from each glucose molecule efficiently in order to make as many ATP molecules as possible. The multiple steps of cellular respiration are described in your textbook. However, our work in this laboratory will be looking at anaerobic cellular respiration in yeast.

Yeast are unicellular microorganisms of the fungi kingdom. They are facultative anaerobes, which means that they can respire or ferment depending on environmental conditions. In the presence of oxygen, respiration takes place (aerobic respiration). Without oxygen, fermentation occurs (anaerobic respiration). Both processes require sugar to produce cellular energy. Here is the chemical reaction of fermentation, which produces ethanol and carbon dioxide as metabolic waste products.

$$\underset{\text{glucose}}{C_6H_{12}O_6} \rightarrow \underset{\text{ethanol}}{2C_2H_6OH} + \underset{\text{carbon dioxide}}{2CO_2}$$

In this laboratory, students will use the respiration powers of yeast to inflate balloons. The purpose of this activity, then, is to reinforce the basic principles of respiration as a fundamental metabolic process for living organisms using yeast as a model. It will also explore how humans use this biological knowledge in everyday life.

Materials

1. Balloons
2. Narrow funnel
3. 1 tablespoon (15 ml) active dry yeast
4. 1 teaspoon (5 ml) sugar
5. measuring spoons
6. measuring cup
7. warm water
8. ruler

Procedure

1. Place the bottom of a funnel into the opening of the balloon. You may need to stretch the opening of the balloon a little bit so that it fits.
2. Pour the yeast and the sugar into the balloon through the funnel. Then fill the measuring cup with warm water from the sink and carefully pour the water into the balloon.
3. Remove the funnel from the opening of the balloon. Tie a knot in the balloon to keep the water-and-yeast mixture inside. Measure your balloon.
4. Place the balloon in a warm place and wait. Measure your balloon again.
5. Now sit back and wait as the balloon gets bigger and bigger.

Discussion

1. What are the reactants in these observed reactions?

   ____

   ____

2. What are the products?

   ____

   ____

3. What is the purpose of warm water and why does it produce the observations it does?

   ____

   ____

4. Why is respiration important for living organisms?

   ____

   ____

5. List what things can you make with yeast.

   ____

   ____

Source: Modified version of an education.com website activity, entitled "Experiment with Balloon Science!"

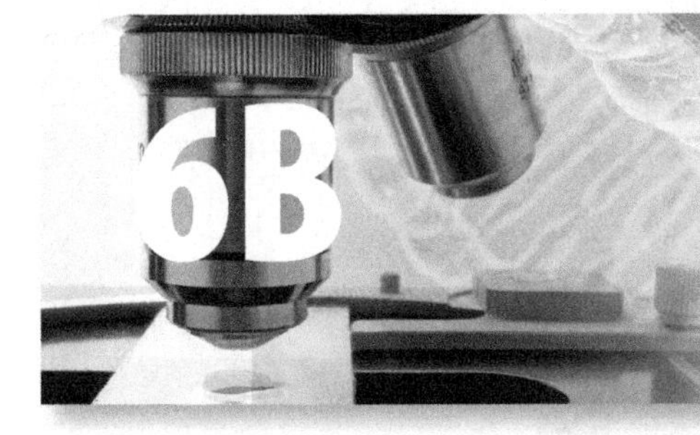

PHOTOSYNTHESIS

The primary source of energy for nearly all life is the sun. The energy in sunlight is introduced into the biosphere by a process known as photosynthesis, which occurs in plants, algae, and some types of bacteria. Photosynthesis is defined as a set of chemical processes by which photosynthetic organisms use water, carbon dioxide, and light energy to drive the synthesis of organic compounds (carbohydrates) and oxygen (gas).

Objectives

To be able to

1. Explain the process of photosynthesis
2. Explain the effects of environmental factors, such as light intensity, temperature, or toxins on the rates of photosynthesis and
3. Explain and calculate Rf values and what they have to do with photosynthesis.

Background

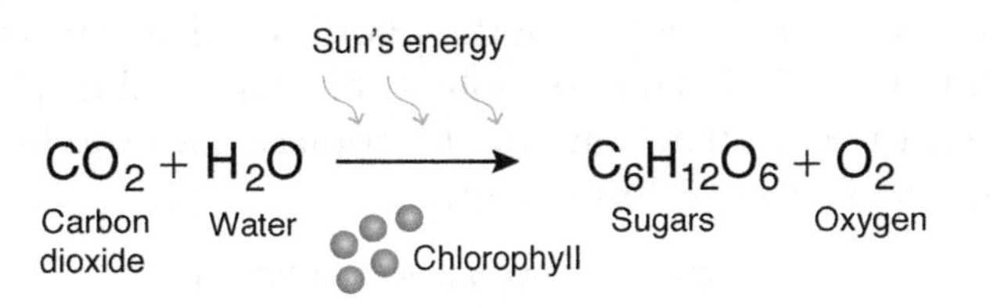

Green plants convert energy in the form of sunlight into energy in the form of biochemical. The wet of processes by which they do this is called photosynthesis. Photosynthesis can happen in plants because they have chlorophyll, which is the pigment that makes plants green. Chlorophyll captures the sun's energy and uses it to make sugars out of carbon dioxide from the air and water. The sugars fuel a plant's roots, stems, and leaves so the plant can grow. During photosynthesis, plants release oxygen into the air. This oxygen is then used by humans and animals to breathe. Make sure you thank plants for the oxygen you use every day!

Chlorophylls are found inside small organelles called chloroplasts. The parts of a chloroplast include the outer and inner membranes, intermembrane space, stroma, and thylakoids stacked in grana. The chlorophyll is built into the membranes of the thylakoids.

Chlorophyll looks green because it absorbs red and blue light, making these colors unavailable to be seen by our eyes. It is the green light which is NOT absorbed that finally reaches our eyes, making chlorophyll appear green. However, it is the energy from the absorbed red and blue light that is, thereby, able to be used to do photosynthesis. The green light we can see is not/cannot be absorbed by the plant and thus cannot be used to do photosynthesis. However, we will learn in one of our experiments today that the chloroplast contains other pigment colors that contribute to photosynthesis.

The Acquisition of Energy: Photosynthesis

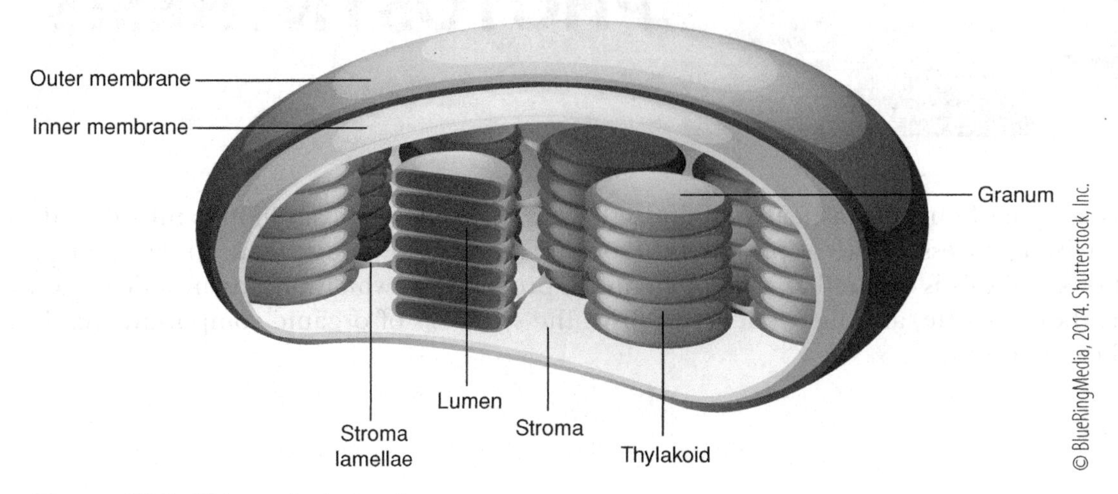

Figure 6B.1 Chloroplast structure.

Chloroplasts are found primarily in plant leaves, and little to none occur in stems. The parts of a typical leaf include the upper and lower epidermis, the mesophyll, the vascular bundle(s) (veins), and the stoma. The upper and lower epidermal cells do not have chloroplasts, thus photosynthesis does not occur there. They serve primarily as protection for the rest of the leaf. The stomas are holes which occur primarily in the lower epidermis and are for air exchange: they let CO_2 in and O_2 out. The vascular bundles or veins in a leaf are part of the plant's transportation system, moving water and nutrients around the plant as needed. The mesophyll cells have chloroplasts and this is where photosynthesis occurs. Between the cells that contain chloroplasts and the lower epidermis is gas space, where CO_2 diffuses into the leaf and O2 diffuses out.

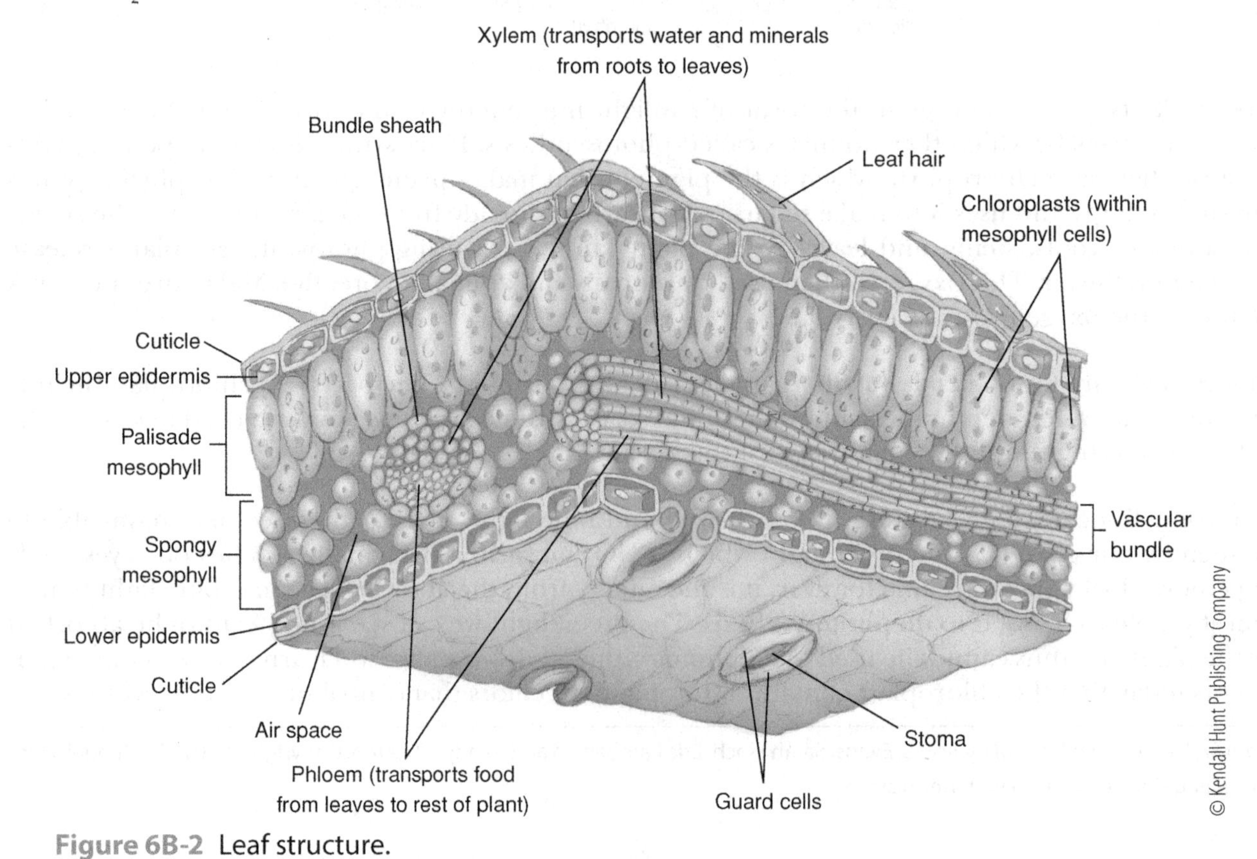

Figure 6B-2 Leaf structure.

ACTIVITY 1: The Photosynthetic Reactions

(The following is based on procedures using Wisconsin Fast Plants.)

How can we measure the rate of photosynthesis? Looking at the overall reaction above, it appears that the easiest measurements would be of the two gases: uptake of CO_2 or production of O_2. We will use a very low-tech approach, taking advantage of the structure of leaves, and measure O_2 production. If we evacuate all the gas from leaf tissue, by pulling a vacuum in it, we can then measure the rate of photosynthesis by observing how quickly the leaf again fills up with gas—the gas O_2 that is produced by photosynthesis. To be able to make this observation, we will carry out the experiment using leaf disks in water. Fresh leaf tissue which has been photosynthesizing will float when placed in water because the space between the lower epidermis and the chloroplast-containing cells is filled with gas. When we evacuate the gas, the leaf disks will no longer float; they will drop to the bottom of the water. As photosynthesis occurs, the space will again fill with gas (O_2), and the leaves will again float. The rate at which the leaf disks float will be our measure of the rate of photosynthesis.

Materials

1. Three- or four-day-old fast plants seedlings
2. Baking soda
3. Small straw
4. 35 mm film can or other accessible container
5. 5 ml syringe

Procedure

1. Add just enough baking soda to just cover the bottom of a small container. A film container will work well, but a 5-ml beaker wrapped in foil will work just as well. Fill the container with water, cover and shake to dissolve the baking soda.
2. Using a straw, cut four leaf disks from the cotyledons of three- to four-day-old fast plants.
3. Remove the cap from the syringe. Next, pull the plunger out of the syringe, then blow the leaf disks out of the straw and into the syringe. Replace the plunger.
4. Draw 4 ml of baking soda solution into the syringe. Invert the syringe with the tip end up. Gently push the plunger to remove all of the air.
5. Next, put your finger over the syringe tip and pull the plunger to create a vacuum, which will pull the air and oxygen from the leaf disks.
6. Tip the end of the syringe down so that the leaf disks are in the baking soda solution. Release the plunger and remove your finger. Turn the syringe back up and tap the side repeatedly until all the disks sink.
7. Place the syringe with the tip up about 5 cm from a light bank or in bright sunlight and record the starting time. You may want to tap the syringe with your finger every 20–30 seconds to dislodge the floating disks.
8. As the leaf disks photosynthesize and produce oxygen, they will float to the top. Record the time at which each disk floats.
9. 9. After all the discs have floated to the top, put the syringe in a dark room or cover with foil. The leaf discs will sink as they respire and consume oxygen.
10. Record the time at which each disc sinks.

NOTE: variations on this experiment that may be used to compare the rate of reactions are:

a. Vary the distance of the light source to the syringe.
b. Place different colors of filters to change the color of the light and how that would affect the rate of the photosynthetic reactions.
c. Add different toxins (in small amounts) to the solution and measure the rate of reaction.

ACTIVITY 2: Plant Pigments

Using Paper Chromatography to Determine the Pigments Present in Leaf Tissue

Chromatography is a process used to separate mixtures of substances into their components. There are various forms of chromatography; however, they all work using the same principles.

All forms of chromatography have a **stationary phase** (a solid or a liquid supported on a solid) and a **mobile phase** (a liquid or a gas). The mobile phase flows through the stationary phase and carries the components of the mixture with it. Different components in the test mixture will travel at different rates, according to a number of factors, such as molecular weight, size of the molecules, type of solvent in the mobile phase, and charge of the molecules on the stationary phase and mixture. In our procedure today, we will use paper chromatography. In this case, the stationary phase is a very uniform absorbent paper and the mobile phase is a suitable liquid solvent or a mixture of solvents.

The distance traveled by a particular compound in a mixture can then be used to identify the compound, based on the mobile phase and the mixture. In this case, we will be separating the green pigment in plant tissue into its component colors. We will do this by calculating the ratio of the distance traveled by a compound to that of the solvent front. This is known as the Rf value; unknown compounds may be identified by comparing their Rfs to the Rfs of known standards.

Rf equation is as follows:

$$Rf = \frac{\text{distance travelled by the compound from the origin}}{\text{distance travelled by the solvent from the origin}}$$

Materials

1. Plant leaves (Fresh young growing plant leaves usually work the best.)
2. A coin with knurled edges (e.g., a quarter or a dime will work best)
3. Chromatography paper cut into strips approximately 1" wide and about 7" long (We do not want to transfer oils from our skin onto the paper so please handle the paper by its edges.)
4. Solvent (Any number of solvents will do. It is recommended to use nine parts ether/one part acetone. Each group will get about 5 ml of the solvent solution.)

Procedure

1. On your strip of chromatography paper, draw a horizontal line with a pencil (not pen) about half an inch from the bottom.
2. Place a plant leaf on the line and roll a coin over it so that you get a line of green pigment on the paper. Using a different part of the leaf, roll the coin again over the same line. Repeat this process until the line is fairly dark.

3. Put about an inch of solvent in a beaker or solvent tank. It is very important that the bottom of the chromatography paper is in the solvent, but the green line is not in the liquid. If the solvent touches the spot directly, the pigment will just dissolve away.
4. The solvent will move up the filter paper slowly and deposit the different pigment components along the way.
5. When the solvent line gets to within about a ¼ inch of the top of the chromatography strip, mark the stopping point (solvent front) with a pencil and allow the strip to air dry.
6. At this point, you should be able to see different colors of pigment on the paper and you are ready to calculate the Rf values for your pigment colors.
   a. First, measure the distance each band travels. This is done by taking a ruler and measuring from the original line to the top of the band.
   b. Next, divide that number by the distance the solvent traveled. This is done by taking your ruler and measuring from the bottom of the strip to the point where the solvent front ends (the pencil line you placed on the chromatography strip when you removed it from the solvent).
   c. Record the pigment color, distance your pigment traveled, the distance the solvent traveled, and the Rf values in the table.

Pigment Color	Distance Pigment Traveled (cm)	Distance Solvent Traveled (cm)	Rf Value
1.			
2.			
3.			
4.			
5.			

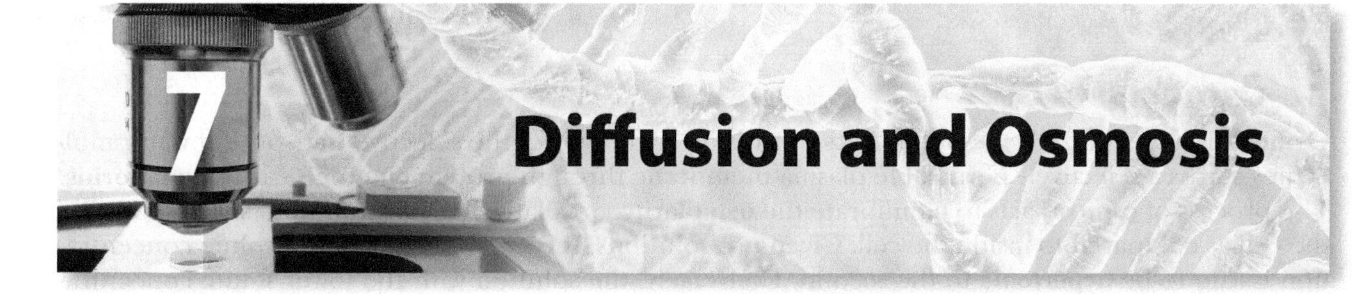

7 Diffusion and Osmosis

Objectives

To be able to

1. Comprehend the concepts of diffusion, osmosis, and selective permeability.
2. Distinguish between hypertonic, isotonic, and hypotonic solutions.
3. Apply the concepts of diffusion and osmosis to living systems.
4. Determine the rate of a chemical process.

A. Diffusion

Diffusion is defined as the spontaneous movement of solutes down their concentration gradient from an area of higher concentration to an area of lower concentration. It is the process of diffusion that drives the basic principles that allow life. The example illustrated in Figure 7.1 depicts free movement of solutes as they diffuse across a membrane. However, the plasma membrane of a living cell is actually **selectively permeable,** or semi-permeable, to many solutes. For example, small nonpolar molecules such as hydrocarbons, CO_2, and O_2 can move freely through the hydrophobic center of the phospholipid membrane, but ions and polar molecules cannot. Cellular gates or channels embedded within the plasma membrane help to control the movement of such solutes into and out of the cell. By controlling what can and cannot move through the cell membrane, selective permeability leads to specialization for different types of cells.

The diffusion of solutes across a plasma membrane without the expenditure of energy is called **passive transport** and can occur with or without the aid of proteins. An example of this is **facilitated diffusion**, in which protein channels that span the plasma membrane allow ions and small polar molecules to bypass the hydrophobic portion of the phospholipid bilayer and diffuse into or out of the cell. Diffusion that requires the input of energy is referred to as **active transport**, in which solutes move against their concentration gradient from an area of lower concentration to an area of higher concentration.

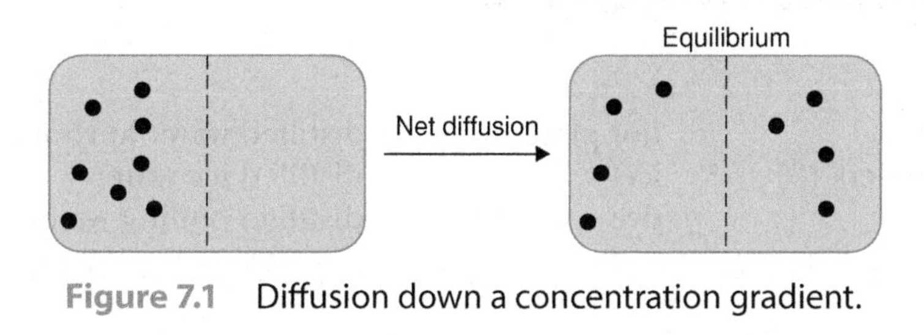

Figure 7.1 Diffusion down a concentration gradient.

B. Osmosis

Osmosis is a specialized type of facilitated diffusion that involves the selective transport of water molecules across a selectively permeable plasma membrane through protein channels called **aquaporins**. The process of osmosis acts to equilibrate the **osmolarity**, or relative concentration of solutes, outside the cell as compared to inside the cell. Given two solutions, the one with the higher solute concentration is said to be **hypertonic** to the second. Conversely, the solution with the lower solute concentration is **hypotonic** relative to the first. Two solutions that have the same concentration of solutes are at equilibrium and are said to be **isotonic** to one another.

Note that because dissolved solutes take up space in solution, a hypertonic solution will contain a lower concentration of water molecules relative to a hypotonic solution. Although water always continues to transport back and forth across a membrane in both directions, the net movement of water molecules will be from the hypotonic side to the hypertonic side, down the concentration gradient of water molecules, but against the concentration gradient of dissolved solutes.

C. Diffusion and Osmosis in Living Organisms

In all living cells, the properties of osmosis are necessary to maintain proper **osmotic pressure**, the pressure exerted on the cell membrane by the flow of water. Excess water movement into a cell will cause it to swell, whereas excess water movement out of a cell will cause it to shrink. Because either situation is potentially fatal to a cell, organisms have mechanisms that regulate water balance. For example, contractile vacuoles in unicellular fresh water protists serve to pump out excess water, and your kidneys function to remove excess water from your entire body. This type of water balance control is known as **osmoregulation**.

The principles that underlie the processes of diffusion and osmosis ultimately determine how individual living cells interact with their surrounding environment, which in turn influences the limits of cellular growth. For example, the transport of O_2 and glucose into a cell is required for cellular respiration, but is limited to the distance that these molecules can diffuse once inside the cell. Therefore, the surface area to volume ratio of an individual cell is essential for ensuring that solutes and molecules are equally distributed throughout the interior of the cell. Because the surface area to volume ratio of many small cells is far higher than that of one giant cell, the limits of diffusion compel cell division and multicellularity.

In this lab, you will examine the properties of diffusion through different kinds of substances, the biological significance of selective permeability of the plasma membrane, and how different cell types respond to changes in solute concentration.

ACTIVITY 1: Diffusion in a Liquid

Materials

Celsius thermometer	hot plate	distilled water at room temperature
15 mL graduated cylinders (3)	ice	distilled ice water
stopwatch	dye	distilled boiling water

Procedure

1. Add 10 mL of cold water to one cylinder, 10 mL of room temperature water to a second cylinder, and 10 mL of hot water to the third.
2. Record the temperature of each water sample in °C.
3. To each cylinder, add 1 drop of dye to the water. Do not mix the dye and water. Let the cylinders sit and observe the diffusion of the dye in each.
4. Use Table 7.1 to record the time in seconds that it takes for the water to turn the color of the dye uniformly.

Results

Table 7.1 Diffusion Time of a Liquid

Solution	Temp. (°C)	Time (sec.)
Cold		
Room temperature		
Hot		

ACTIVITY 2: Diffusion in a Solid

Materials

agar plates
forceps
metric ruler
methylene blue crystals (MB)
potassium permanganate crystals

Procedure

1. Obtain one agar plate.
2. Place equal amounts of methylene blue ($C_{16}H_{18}N_3SCl$) crystals and potassium permanganate ($KMnO_4$) crystals approximately 70 mm apart on the agar surface and 10 mm from opposite edges of the plate as shown in Figure 7.2. Be sure to note which dye is which.
3. Use a metric ruler to measure the distance in millimeters (mm) that each dye diffuses at 15 minute intervals over a period of 90 minutes.
4. Record your results in Table 7.2.

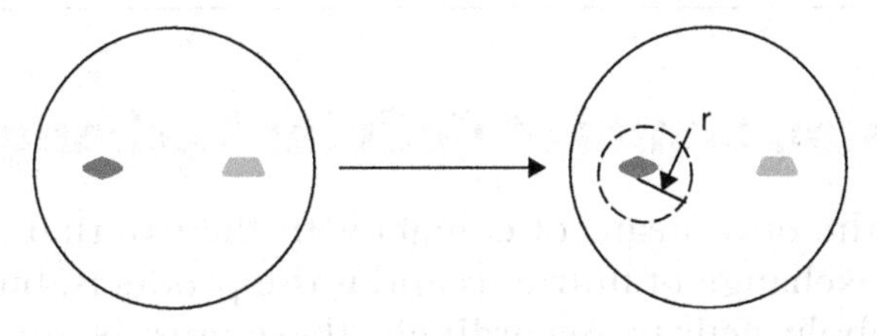

Figure 7.2 Diffusion of a solid.

Results

Table 7.2 Diffusion Distance of a Solid

Time (min)	MB (mm)	KMnO (mm)
15		
30		
45		
60		
75		
90		

ACTIVITY 3: Diffusion in a Gas

Materials

ammonium
meter stick
stopwatch

Procedure

1. Using the meter stick, measure the distance in meters (m) from the front of the room to your lab bench. Record this distance below.
2. Your instructor will open a bottle containing ammonium.
3. Using the stopwatch, determine the time in seconds (sec) that it takes for you to smell the ammonium. Record this time below.
4. Using your distance and time measurements, calculate the rate of diffusion (m/sec) of a gas in air.

Results

Distance from front of room to your lab bench (m) = ______________________

Time to your lab bench (sec) = ______________________

Diffusion rate (m/s) = ______________________

ACTIVITY 4: Diffusion Limits of Cellular Exchange

The cellular membrane is the only means of contact with the external environment. This interface not only allows a constant exchange of nutrients and waste products, but it regulates the passage of substances critical for metabolic activity. Accordingly, there must be enough surface area to accommodate the increasing volume within a growing cell. This exercise will demonstrate how smaller cells are more efficient than larger ones in handling increased cellular volume.

Materials

metric ruler
beaker
spoon
scalpel blade
paper towels
0.1 M NaOH
agar blocks of various sizes prepared with phenolphthalein

Procedure

1. Record all measurements in Table 7.3 below.
2. Using a metric ruler, measure the length (l) in millimeters (mm) of one side of each agar block.
3. Calculate the surface area (mm^2) and volume (mm^3) of each cube using the following formulas:

$$\text{Surface Area (S) of a cube} = 6 \times \text{length squared } (l^2)$$
$$\text{Volume (V) of a cube} = \text{length cubed } (l^3)$$

4. Calculate surface area to volume ratio (S/V). Record your answer.
5. Soak each cube in a beaker containing 0.1 M NaOH for 5 minutes.
6. Remove the agar cubes with a spoon and blot them dry with a paper towel.
7. Refer to Figure 7.3. On one side of each cube, mark the halfway point (dotted line). Measure the *half-length* distance (A) in mm from the halfway point to the edge of the cube. Record your measurements.
8. With your scalpel blade, cut each agar cube in half. Measure the depth of the pink-colored area (B) in mm and record it as *movement*. Calculate *movement* to *half-length ratio* (B/A). Record your measurements.
9. Use your measurements to construct a graph in Figure 7.4 showing B/A (movement/half-length ratio) as a function of S/V (surface area/volume ratio).

Results

Table 7.3 Diffusion-Limited Exchange

Length (L)	Surface area (S)	Volume (V)	S/V ratio	Half-length (A)	Movement (B)	B/A ratio

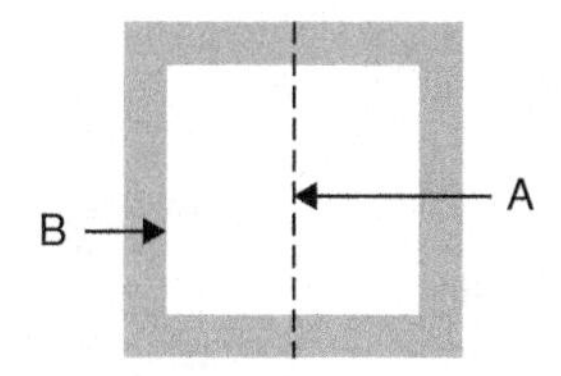

Figure 7.3 Diagram of an NaOH-soaked agar cube. A is the halflength of the cube; B is the distance moved by the NaOH.

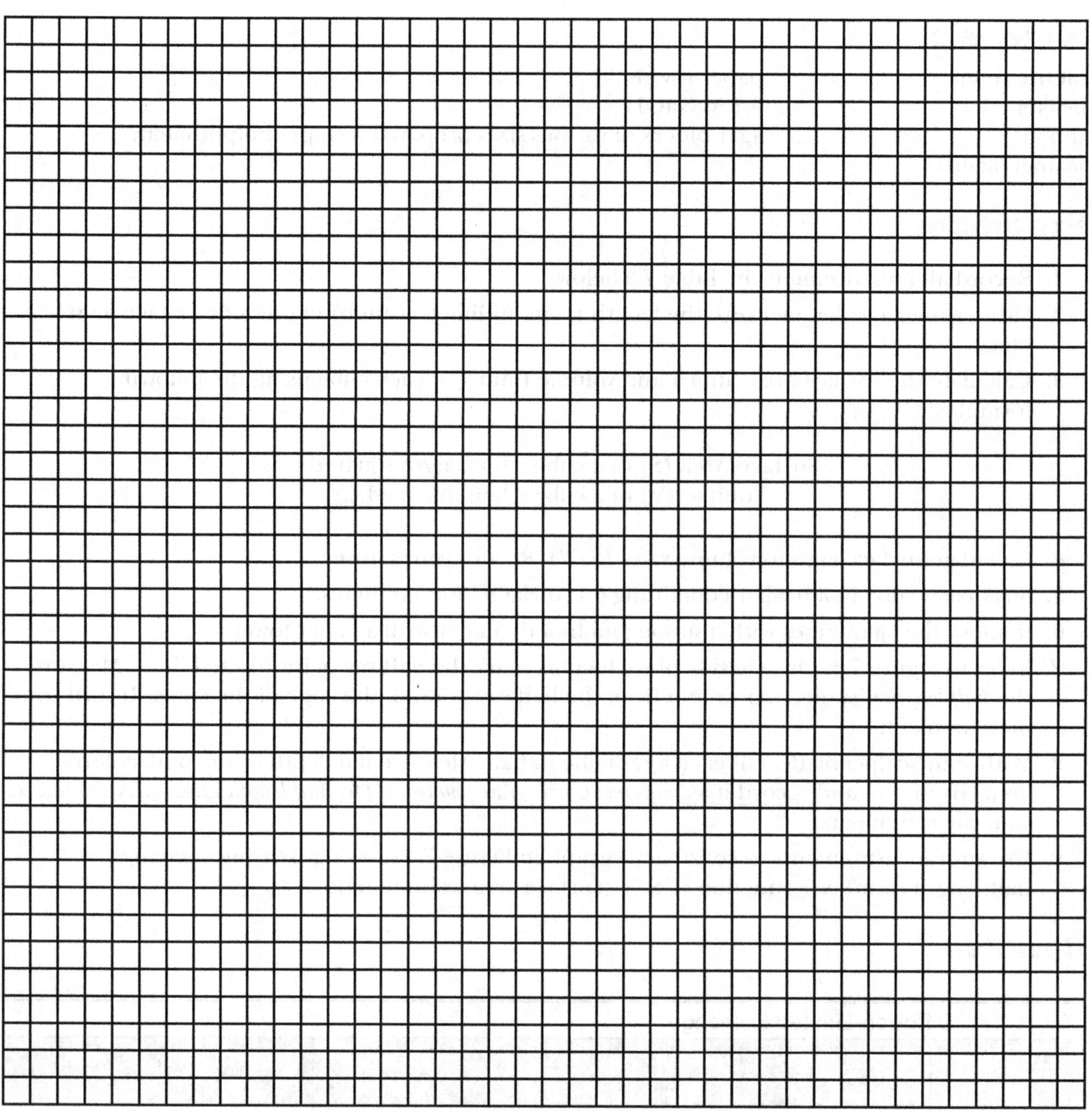

Figure 7.4 Effect of increasing size of an object on the rate of diffusion.

ACTIVITY 5: Osmosis through a Selectively Permeable Membrane

Materials

dialysis tubing
twine
150 mL beakers (2)
test tubes (6)
test tube rack
marking pencil
pipettes
Benedict's solution
balance
40% maltose solution
distilled water

Procedure

1. Obtain four pieces of dialysis tubing, each 15 cm long, and presoak in distilled water.
2. For each piece of tubing, fold over one end. Tie the folded end securely with twine. This will make the tubing into a bag.
3. To two of the dialysis bags, insert 15 mL of 40% maltose solution into the open end.
4. To the other two bags, insert 15 mL of distilled water into the open end.
5. Fold and tie off the open end of each bag as in step 2, leaving only a small amount of air in each bag.
6. Observe the color of the solution within each bag. Record the weight in grams (g) of each bag and record in Table 7.4 in the column labeled "0 min." Rinse the outside of the bags with distilled water.
7. Add 100 mL of distilled water to a 250 mL beaker and add 100 mL of 40% maltose solution to another 250 mL beaker.
8. Into the beaker containing water, immerse one bag containing distilled water and one bag containing 40% maltose. Mark each bag clearly to indicate its contents. Make sure each bag has been rinsed before proceeding.
9. Into the beaker containing 40% maltose, immerse one bag containing distilled water and one bag containing 40% maltose.
10. After 10 minutes, remove each bag from its beaker, dry lightly on a paper towel, and weigh the bag in grams. After weighing, replace each bag in its beaker. Record the weight in Table 7.4.
11. Repeat step 10 at 10, 20, 30, 40 and 50 minutes. Record each weight in Table 7.4.
12. After the fi nal weighing at 50 minutes, use a pipette to remove 1 mL of solution from each dialysis bag and transfer to labeled test tubes. (Be careful not to spill any contents of the dialysis bags into the beakers.)
13. To each test tube, add 5 mL of Benedict's solution and place in a boiling water bath for 5 minutes. Record the color of any precipitate that forms in Table 7.4.
14. Use the data in Table 7.4 to construct a graph in Figure 7.5. Put time (min) on the x-axis and weight (g) on the y-axis. Label each set of points and determine the rate of osmosis for each sample.

Results

Table 7.4 Osmosis through Selectively Permeable Membrane

Conditions	0 min	10 min	20 min	30 min	40 min	50 min	Color of bag contents after Benedict's test
Water in bag; Maltose in beaker							
Water in bag; Water in beaker							
Maltose in bag; Water in beaker							
Maltose in bag; Maltose in beaker							

Figure 7.5 Rate of osmosis of 40% maltose through a selectively permeable membrane.

ACTIVITIES TO REINFORCE CONCEPTS

ACTIVITY 6: Osmosis through a Living Animal Membrane

Materials

light microscope
microscope slides (3)
coverslips
disposable gloves
animal blood
distilled water (dH_2O)
saline (0.9% NaCl)
15% NaCl solution

Procedure

1. Place 1 drop of animal blood on each of the three microscope slides.
2. To the first drop of blood, add one drop water; to the second drop, add one drop of saline; and to the third drop, add one drop of 15% NaCl.
3. After 5 minutes, place a coverslip over each sample to prepare wet mounts of the red blood cells.
4. Observe the red blood cells under the 40x objective lens of the microscope.
5. In Figure 7.6, draw and describe what you observe on each slide.
6. Discard the slides in an appropriate container.

Results

Water	Saline	15% NaCl

Figure 7.6 Animal blood cells.

ACTIVITY 7: Osmosis through a Living Plant Membrane

Materials

light microscope
microscope slides (3)
coverslips
Elodea
distilled water (dH_2O)
saline (0.9% NaCl)
15% NaCl solution

Procedure

1. Place one leaf of *Elodea* on each of the three microscope slides.
2. To the first leaf sample, add one drop water; to the second sample, add one drop of saline; and to the third sample, add one drop of 15% NaCl.
3. After 5 minutes, place a coverslip over each sample to prepare wet mounts of the *Elodea* leaves.
4. Observe the *Elodea* leaf cells under the 40x objective lens of the microscope.
5. In Figure 7.7, draw and describe what you observe on each slide.
6. Discard the slides in an appropriate container.

Results

Water	Saline	15% NaCl

Figure 7.7 *Elodea* cells.

ACTIVITY 8: Effect of Osmosis on a Living Organism

Materials

dissecting microscope
microscope slides (3)
coverslips
flatworm (in pond water)
distilled water (dH_2O)
pond water
15% NaCl solution

Procedure

1. Obtain a flatworm in pond water and place it on a microscope slide.
2. Observe the flatworm under the dissecting scope. In Figure 7.8, draw and describe what you observe.
3. Carefully drain the pond water with a piece of paper towel and add one drop of 15% NaCl.
4. Observe the flatworm for several minutes. Draw and describe what you observe.
5. Carefully drain the NaCl with a piece of paper towel and add one drop of distilled water.
6. Observe the flatworm for several minutes. Draw and describe what you observe.
7. Discard the slides in an appropriate container.

Results

Water	Saline	15% NaCl

Figure 7.8 Flatworm.

Laboratory Report Sheet

NAME______________________________ **SECTION** __________ **GRADE** __________

Diffusion and Osmosis

1. How does temperature affect the rate of diffusion?

2. In your osmosis experiment, what do the results of the Benedict's test tell you?

3. Explain what would happen to each of the following cells if they were placed into a solution that is hypotonic, hypertonic, or isotonic:
   a. red blood cell –

   b. a cell from a spinach leaf –

   c. a fresh water amoeba –

4. You have two dialysis bags that are permeable to water and to NaCl, but not permeable to sucrose. You fill each bag with 10% NaCl and seal each end tight. One bag is to be placed in a solution of 10% NaCl. The second bag is to be placed into a solution of 10% NaCl and 30% sucrose. From your knowledge of diffusion and osmosis, explain what will happen to each as they sit in these solutions.

8A DNA Structure and Function

What's the Point?

Biological Anthropology has experienced something of a revolution in the past several decades. Since the discovery of DNA, scientists have been able to learn things about ourselves and our ancestors that were previously unknown or thought unlikely to be related to genetics (e.g., relationship of genes to behavior and disease). The application of knowledge of DNA is not limited to human evolution, either; developments have been made in food production, medicine, and even law enforcement, changing the lives of billions of people.

In the first week, we established that the concept of individual variation, as important to survival and reproduction, began with Darwin. No early scientific thinkers, even Darwin, had any knowledge of genes or DNA. Darwin thought that traits must be heritable, but he did not know what was transmitted between generations or how.

Before we can discuss how genetics enhances our understanding of evolution, we should step back and examine the basics.

The Discovery of DNA

The discovery of DNA, **deoxyribonucleic acid,**[1] in 1953 by Francis Crick, James Watson, and Rosalind Franklin followed a long process of important scientific discoveries. It was already known that heredity, or the passing of traits from parent to offspring, is central to our definition of life. Gregor Mendel's work (circa 1856) showed us that variants of genes are expressed as visible traits or **phenotypes.** Thomas Hunt Morgan's work on the genetics of fruit flies in the early 1900s demonstrated that genes resided on something called **chromosomes**. Yet, it still was not clear to scientists what the actual inherited genetic material was. What are genes, or even chromosomes, made of? There are many different kinds of molecules present in an **organism**, but which one was passed between generations? Further experimentation on bacteria by Oswald T. Avery in the 1940s gave strong evidence that the molecule that carries heritable information is DNA.

> If we are right, and of course that is not yet proven, then it means that nucleic acids are not merely structurally important but functionally active substances in determining the biochemical activities and specific characteristics of cells and that by means of a known chemical substance it is possible to induce predictable and hereditary changes in cells. This is something that has long been the dreams of geneticists—Oswald T. Avery, 1943[2]

[1] Find definitions of words in the glossary at the end of this primer. Definitions were taken, all or in part, from Mai, L.L., Young Owl, M., and Kersting, M.P. 2005. *The Cambridge Dictionary of Human Biology and Evolution.* Cambridge UK: Cambridge University Press. Quotes are so indicated.

[2] http://profiles.nlm.nih.gov/ps/retrieve/Narrative/CC/p-nid/157

The next logical question that the scientific community posed was of course, "how?" How does DNA function as the carrier for genetic information? The answer to that question lies in the structure of DNA itself.

Chemical Composition of Nucleic Acids

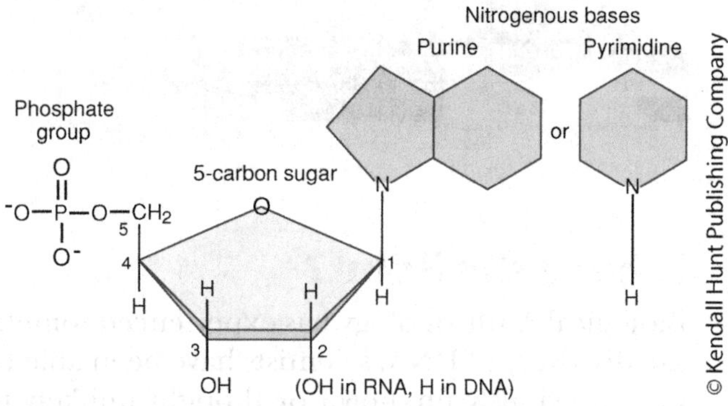

Figure 8A.1 Two types of bases, double-ringed purines and single-ringed pyrimidines, are attached to a sugar-phosphate group to form a nucleotide.

Scientists in the 1940s already had a firm understanding of the components that made up DNA and other **nucleic acids**. They knew that there were four nitrogenous (nitrogen-containing) bases: Adenine (A), Cytosine (C), Guanine (G), and Thymine (T). They also knew that there was lots of phosphoric acid and a little deoxyribose sugar in every DNA molecule. Scientists also noticed that As and Gs looked structurally similar to purines (a double-ringed chemical compound: adenine and guanine, Figure 8A.1), while Cs and Ts looked similar to pyrimidines (a single-ringed compound).

A subunit of DNA is complete when a phosphate group is attached to a sugar ring and a **base** through a strong chemical bond. These structures are called **nucleotides** (see Figure 8A.1). Nucleotides then attach to each other along both the "rungs" of the ladder via hydrogen bonds and along the phosphate backbone to form a DNA molecule (Figure 8A.2)

The Human Genome

Every cell in your body contains a complete copy of your DNA, called a **genome**. Throughout the entire animal kingdom, genome sizes range from 159,662 bases of Carsonella ruddii (a bacterium that lives in sap-feeding insects) to 670 billion bases of Amoeba dubia (a freshwater parasite). Humans fall about in the middle, with approximately 3.2 billion base pairs. The genome of corn is almost the same size as ours at 2.5 billion bases. After considering this, you may begin to realize something that scientists did not appreciate until the early 1970s: genome size does not predict complexity of the organism.

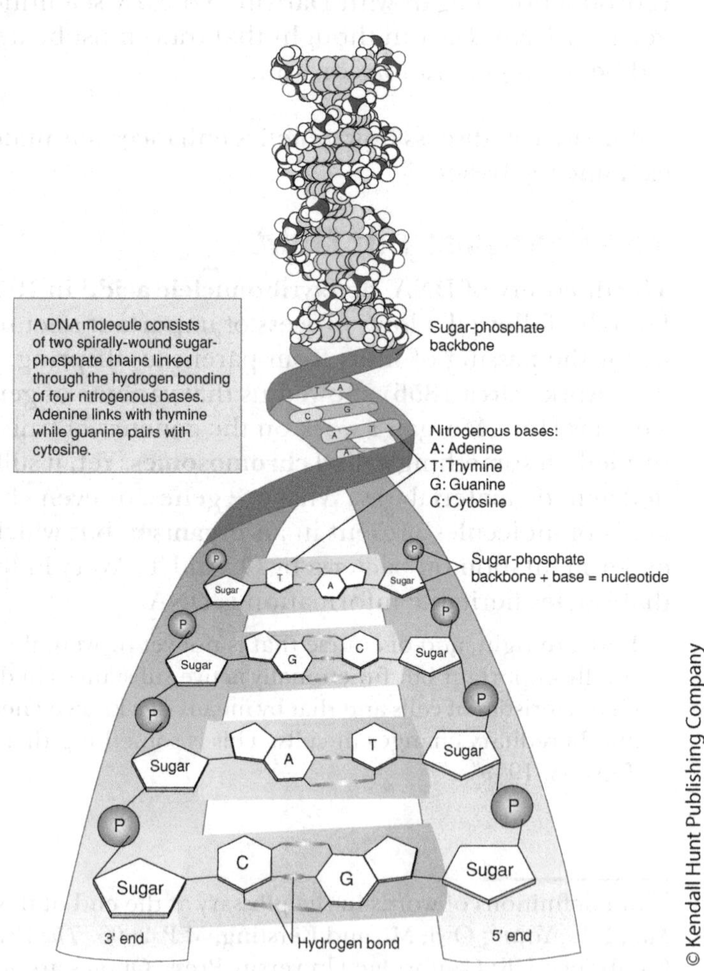

Figure 8A.2 DNA replication (also see text).

The human genome contains about 20,000 genes, which actually only make up slightly less than 2% of our entire genome. Genes are important because these segments of DNA code for different **protein** products, which are chains of **amino acids** that are essential to almost all cellular processes. As mentioned in the first week of the course, we are literally composed of proteins, so their expression can be observed across humans as variable traits or **phenotypes**. The rest of the genome is made up of **noncoding DNA**. Within the noncoding regions of DNA are stretches of DNA that regulate the expression of a trait, called **regulatory DNA**. These sequences basically tell the coding regions, or genes, how and when to be expressed. Just as a conductor dictates which members of an orchestra play when and for how long, a regulatory gene controls the expression of other genes. They tell the genes when to become active, how much protein product to produce, and even which protein products go where. A small number of regulatory regions can affect the expression of genes in different ways, making it possible to generate a great deal of biological complexity with relatively few genes.

How Does DNA Carry Genetic Information?

While DNA only contains four nucleotides (As, Gs, Cs, Ts), they are not simply repeated units. DNA can carry extremely complex information by varying the sequence of bases. A DNA sequence that is two nucleotides (for example, CG, CC, GC, GG, AT, AA, TA, TT) long would have 16 different possible sequences (4 possible bases in the first spot, 4 possible bases in the second spot; $4 \times 4 = 16$). If you have a sequence of bases that is 6 nucleotides long, you would have 4,096 different possible base sequences ($4 \times 4 \times 4 \times 4 \times 4 \times 4 = 4{,}096$). Considering there are over 3 billion base pairs in each human cell, the number of different possible sequences is truly staggering!

In 1953, research scientists James Watson, Francis Crick, and Rosalind Franklin finally discovered the structure of DNA, which revealed the mechanism by which DNA can be faithfully copied, allowing the astonishing amount of information that is stored in DNA to be passed on to the next generation. Using a technique called X-ray diffraction, Rosalind Franklin at King's College, London, took a picture of DNA that revealed an interesting pattern. Upon inspection of the image, Watson and Crick determined that DNA must be *a simple, regular structure with repeated units.*

From the distinctive pattern in the X-ray diffraction photo, Crick deduced that DNA should be a *double helix,* with the phosphate groups on the outside and the bases on the inside. Next, Watson hoped to determine how the nitrogen bases are arranged. In a moment of insight and luck, he realized that A could pair closely with T, and that G could pair closely with C. Moreover, the A/T base pair was about the same *width* as the G/C pair, which is consistent with the fact that a diameter of a helix is *regular for the length of the structure.* On the basis of the angles of certain chemical bonds and the proximity of the base pairs, Crick was able to point out that *the two helices had to run in opposite directions.* The helices are antiparallel to each other.

"We have Discovered the Secret of Life!"

Legend has it that Watson and Crick burst into a nearby pub after they realized what their discoveries meant and proclaimed that they "have discovered the secret of life!" Watson and Crick submitted a 1-page manuscript on the structure of DNA to **nature** in 1953.[3] They described DNA as a twisted ladder with curving, parallel sides (think of a slinky or a twisted telephone cord): the sugar phosphate backbone. The rungs of the ladder represent bases pairing together with hydrogen bonds (see Figure 8A.2). The very configuration of DNA allows it to be faithfully replicated. **Base pairing** allows each strand to serve as a template for making its own complementary partner during **replication**. The result is two new full strands from one original. Proofreading functions in the replication machinery ensure that mismatched bases happen infrequently. Yet mistakes or mutations happen in about 1 in 10 billion matches. Mutations happen infrequently enough that mistakes in our genetic code are not something most people have to worry about (although some mutations can result in cancer), but are frequent enough that over billions of years of evolution, mistakes in DNA replication resulted in the vast diversity of life we see on Earth today.

---

[3]Watson J. D., and F. H. C. Crick. 1953. "Molecular structure of nucleic acids: a structure for deoxyribose nucleic acid." *Nature* 171(4356): 737–38.

DNA Replicates Itself Before Cells Divide

DNA copies itself (replicates) in every cell that divides so that each new cell receives a full copy of the organism's DNA. [We will go into more detail about HOW cells divide next week.] Before cells can divide, DNA condenses into compact, organized structures called chromosomes (Figure 8A.3), which makes the entire genome 10,000 times shorter than uncondensed DNA.

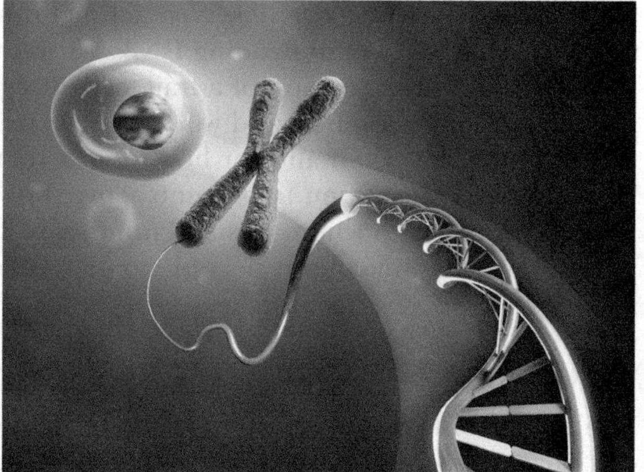

Figure 8A.3 In preparation for cell division, the DNA helix condenses and becomes tightly wound into structures that are visible under a light microscope. In this figure, notice that the DNA double helix has already replicated. The chromosome is "double-stranded." Note that each arm of the chromosome consists of a DNA double helix. The two arms have identical genetic material.

Humans have two sets of 23 chromosomes, for a total of 46 (numbered 1 through 22–somatic chromosomes–plus a pair of sex chromosomes; XX = female and XY = male; Figure 8A.4). Because we inherit one of each chromosome from each parent, we have two sets of chromosomes and are **diploid**. All eukaryotic organisms are diploid, whereas all prokaryotes (cells without a nucleus) are haploid, meaning they only have one set. In eukaryotes, the copies of each chromosome form **homologous pairs**. They are homologous ("same") because they have the same gene loci (plural for locus). Genes occupy a place or locus on the chromosome, but the sequences of bases on the gene may be the same or different in each homologous pair. Homologous pairs of chromosomes may have different (mutated) versions of the same gene, called **alleles** at a particular locus. If different, then one gene or allele was received from one parent and a different allele from the other parent. For instance, if you inherited *two different* hair color genes from each of your parents, one may code for black hair and the other may code for brown hair. One or both of those genes may be expressed in your phenotype (i.e., your physical characteristics).

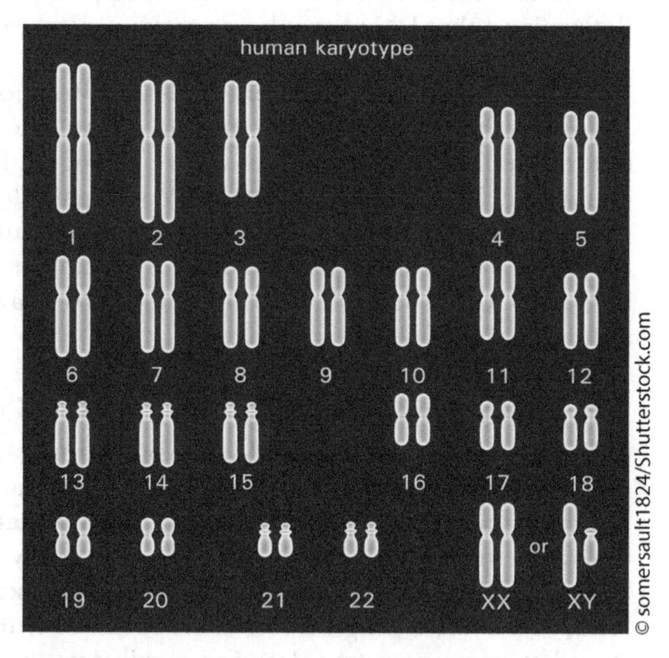

Figure 8A.4 Normal human karyotype. The homologous chromosomes (one from each parent) are arranged side by side. Numbers identify chromosome size and position of centromere (constriction). This karyotype shows both male and female sex chromosomes. All of the other chromosomes (1–22) are called "autosomes."

DNA replication begins as an enzyme splits the hydrogen bonds that keep the two DNA strands together (Figure 8A.5). The DNA molecule is in the process of doubling or replicating the entire sequence of bases or genes. The white strands are the original molecule; the black and white strands represent the replication of the original strand into two identical new strands.

Enzymes "unzip" the hydrogen bonds that bind complementary bases. Free nucleotides containing a base + sugar + phosphate molecules are attracted chemically to the exposed bases. For example, C attracts G and G attracts C, whereas

A attracts T and T attracts A (see Figure 8A.5). As the correct complementary bases bond to exposed bases, the original molecule becomes two new ones. This is occurring from bottom to top. The result is two DNA molecules with the same sequence of bases as the parent strand. DNA has replicated! If the replicated sequence is not identical to the original one, if bases are substituted or deleted, then a mutation (a change in the original base sequence) has occurred.

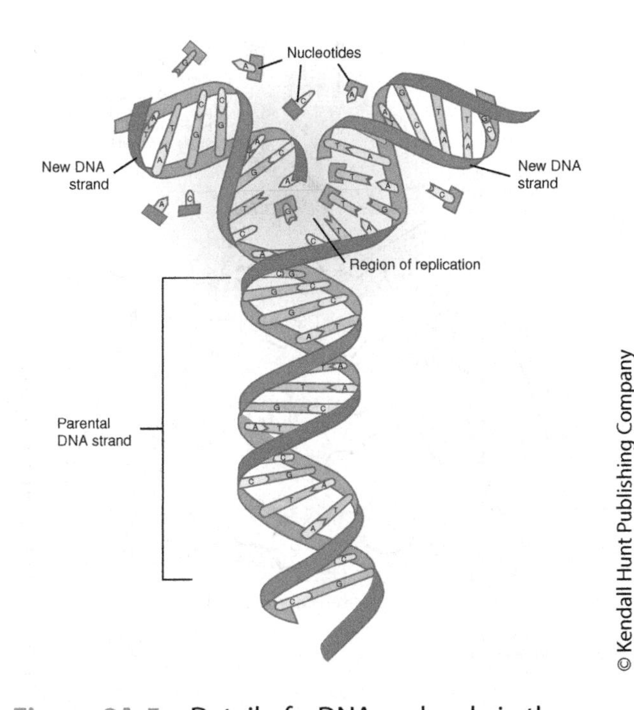

Figure 8A.5 Detail of a DNA molecule in the early stages of replication. The hydrogen bonds have dissolved in the upper part of the molecule, exposing bases. The base portion of free-floating nucleotides (a base: Cs are attracted to Gs (and vice versa); As are attracted to Ts. The new strands are identical to each other and identical to the original molecule. Each new DNA strand will have one original side and one new side.

Gene Expression (Protein Synthesis)

DNA contains the information to build protein products, such as the keratin that builds your fingernails, but DNA itself does not do the building. It is too large to pass through the nuclear membrane into the cytoplasm, where proteins are actually assembled. Instead, it is copied into a smaller format, called **messenger RNA**, which is a subtype of RNA (**ribonucleic acid**). RNA (Figure 8A.6) is similar to DNA, but differs in a few very important ways.

Scientists in the 1940s were able to identify the components of RNA at the same time they identified DNA. They noticed that instead of thymine (T), RNA had a base called *uracil (U)* in its place. They also observed that instead of deoxyribose sugar, RNA has ribose sugar, which is slightly different in its chemical make-up. RNA is also *single stranded and much shorter* than DNA. Instead of copying the entire genome (as occurs before cell division and DNA replication), RNA copies one gene at a time.

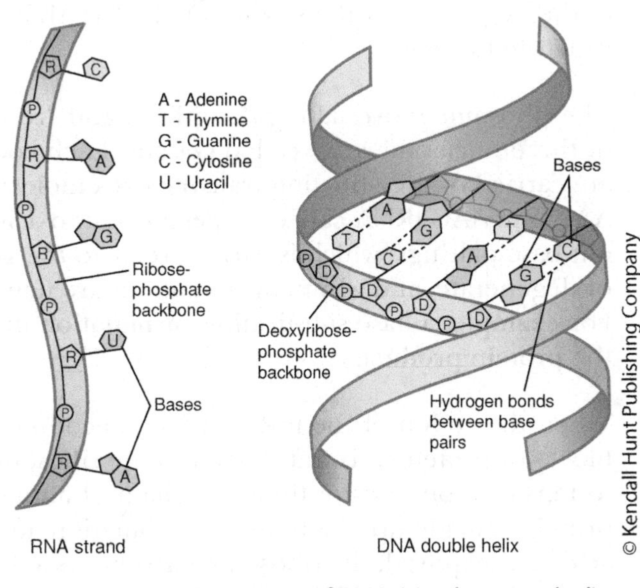

Figure 8A.6 Structure of RNA (single-stranded) compared with DNA (double-stranded).

Proteins are assembled in two major stages, **transcription** (Figure 8A.7) and **translation** (Figure 8A.8). When transcription begins, DNA receives a chemical signal to unwind and separate at a specific gene. Separation of the two DNA strands occurs by breaking the hydrogen bonds down the middle of the molecule (the rungs of the ladder), much like the beginning of DNA replication. Free-floating nucleotide bases in the nucleus are lined up along the template strand by an enzyme called RNA polymerase and are formed into a complementary mRNA strand. Once the segment of DNA bases has been copied, the mRNA detaches and travels through the nuclear membrane into the cytoplasm. Meanwhile, the DNA strand winds itself back up and remains in the nucleus. Once the mRNA is in the cytoplasm, a large complex

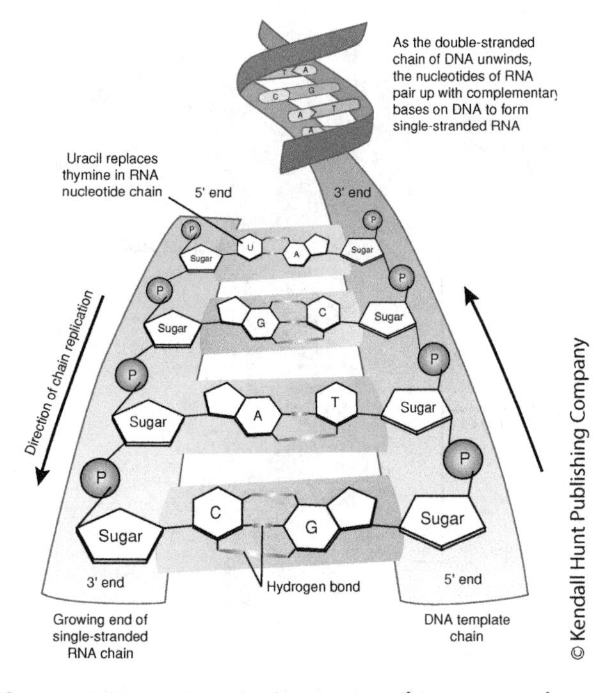

Figure 8A.7 Transcription: making a single-stranded messenger RNA molecule (also see text).

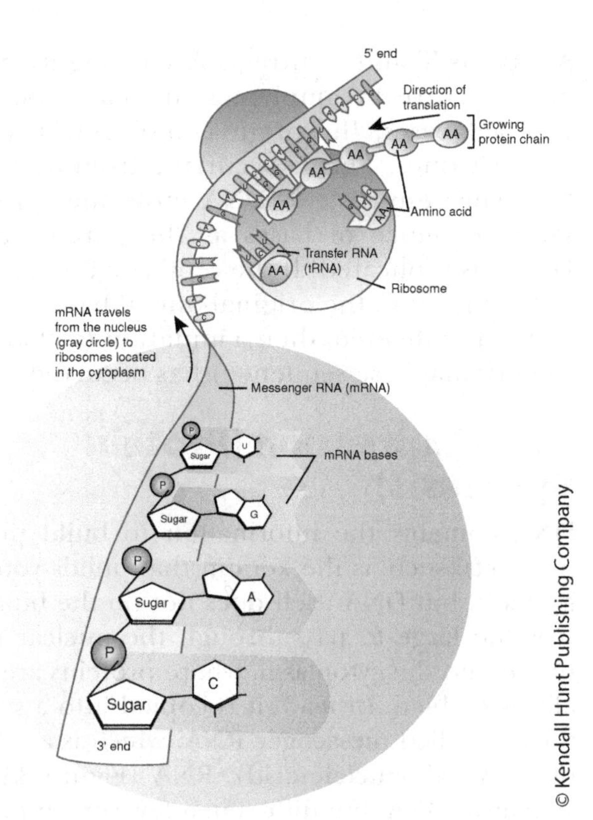

Figure 8A.8 Translation. mRNA is pulled through a ribosome to sequence amino acids and produce a protein.

molecule called a ribosome, which is made up of RNA and many proteins, is recruited to the mRNA at the beginning of the strand, or the "START" codon. This begins the second stage of expression, called *translation.*

The ribosome moves along the mRNA and "reads" the code three bases at a time. These sequences of three nucleotides are called **codons**. Each codon is complimented by a specific **anticodon**, which are carried on free-floating **transfer RNA** molecules (e.g., in Figure 8A.9, mRNA: UUG meets tRNA: AAC; the AAC tRNA carries a specific amino acid). Most amino acids are coded for by several different base pairings, which is why we refer to the **genetic code**(see text for Figure 8A.9) as redundant, or degenerate, and why many mutations are neutral and do not change the protein being assembled. For example, a base substitution or **mutation** in the third letter of a codon usually has no effect on the protein product.

When the codon of the mRNA and anti-codon of the tRNA matches, the amino acid (the building blocks of proteins), is transferred to the ribosome. After the process is repeated (as few as ten or up to tens of thousands of times), a chain of amino acids (a polypeptide, or protein) is formed. Each protein is made up of a unique combination and number of amino acids. Once the required amino acids are acquired, the ribosome encounters a STOP, or nonsense codon, that doesn't correspond to any of the amino acids. Translation is complete and the ribosome and mRNA degrades, leaving the assembled protein to carry out its function.

Transcriptional Regulation and Evodevo

In your body, each one of the several trillions of cells is not the same as all the others; there are numerous *types* of cells, which cooperate with others like them to compose tissues, which themselves are the basis of the many organs with which you are probably already familiar. Those organs cooperate in making the large-scale organ systems, which cooperate in maintaining the whole body. Despite this astronomical complexity, every one of these cells is working with virtually the same set of instructions, the DNA that is housed in the nucleus of each and every one of them.[4] These instructions allow every type of cell to build, or synthesize, each protein that it will need in order to properly function. Above, you learned that the synthesis of proteins occurs via translation and that translation requires the transcription of mRNA. Here, we will approach the questions of how the cell begins, ends, or otherwise regulates the transcription of genes. This knowledge is currently revolutionizing biology and some fields of biological anthropology through a newly emergent subdiscipline, known as evolutionary developmental biology, or *evodevo* for short.

First Base	Second Base: U	Second Base: C	Second Base: A	Second Base: G	Third Base
U	UUU phenylalanine	UCU serine	UAU tyrosine	UGU cysteine	U
U	UUC phenylalanine	UCC serine	UAC tyrosine	UGC cysteine	C
U	UUA leucine	UCA serine	UAA stop	UGA stop	A
U	UUG leucine	UCG serine	UAG stop	UGG tryptophan	G
C	CUU leucine	CCU proline	CAU histidine	CGU arginine	U
C	CUC leucine	CCC proline	CAC histidine	CGC arginine	C
C	CUA leucine	CCA proline	CAA glutamine	CGA arginine	A
C	CUG leucine	CCG proline	CAG glutamine	CGG arginine	G
A	AUU isoleucine	ACU threonine	AAU asparagine	AGU serine	U
A	AUC isoleucine	ACC threonine	AAC asparagine	AGC serine	C
A	AUA isoleucine	ACA threonine	AAA lysine	AGA arginine	A
A	AUG(start) methionine	ACG threonine	AAG lysine	AGG arginine	G
G	GUU valine	GCU alanine	GAU aspartate	GGU glycine	U
G	GUC valine	GCC alanine	GAC aspartate	GGC glycine	C
G	GUA valine	GCA alanine	GAA glutamate	GGA glycine	A
G	GUG valine	GCG alanine	GAG glutamate	GGG glycine	G

Figure 8A.9 The Genetic Code table that identifies the three-letter codes for amino acids. There are four bases that may be arranged in 64 possible sequences. Note that there are several repeats of the amino acids, thus the code is referred to as "redundant." A mutation in one letter may result in no change in the amino acid (or subsequent protein). If the mutation results in a change in the amino acid it codes for, the protein will be different (for better or for worse). Thus, mutations are the source of all new variation in a population.

In order to be eventually translated into a protein, a gene must, of course, be transcribed. Transcription is caused by many cooperating proteins that work directly on the DNA itself. At a bare minimum, transcription requires a **promoter** (also called a "TATA box"), which is a segment of DNA at the beginning of a gene that binds to the proteins that will actually read and transcribe the DNA. Working on the promoter are the RNA polymerase mentioned above and other proteins and sections of DNA called **transcription factors**; the polymerase actually does the transcribing, while the transcription factors help the polymerase to get started. As you probably already expect, though, the promoter is not the whole story. On the same chromosome, sometimes very distant to the gene being transcribed, there are also regions known as **enhancers,** which can bind to transcription-factor proteins in order to regulate the transcription of a gene.

Via regulatory processes, transcription can be either turned off or turned on. If transcription is turned on, then it can be either **up-regulated** or **down-regulated**. For example, the transcription of proteins that allow electrochemical impulses to travel across a cell is not needed in muscle cells, so transcription of these proteins is *turned off* in muscle cells and *turned on* in nerve cells. In some *types* of nerve cells,

[4] Some cells, brain cells in particular, can actually have different genomes than other cells in the body. However, these differences are acquired over the course of development, and it is unknown whether or not these differences have any function (Macosko, EZ, and McCarroll, SA. 2013. Our fallen genomes. *Science,* 342 (6158), 564–565. doi:10.1126/science.1246942)

these proteins are required in greater amounts than in others; therefore, transcription of the genes that code for such proteins will be up-regulated in these cells. As already outlined, this entire process is mediated by other proteins, which must be synthesized through their own pathways of transcription and translation; therefore, even these proteins can be subject to regulation themselves! Thinking through this whole process, you may be wondering how it begins. After all, if a protein requires other proteins in order to be synthesized, how did those first proteins get there? This question pulls us closer to a revolutionary new subdiscipline of biology known as evodevo, or evolutionary developmental biology, which we will discuss in more detail in the next chapter.

Summary

A little more than 50 years ago, the structure of DNA was discovered by Watson, Crick, and Franklin. Since then, a tremendous amount of research has resulted in improved understanding about genetic relatedness between humans and other organisms, and among human populations as well as gene interactions that result in many diseases and as well as normal traits of humans. Interestingly, the understanding that DNA is a code that never leaves the nucleus, but sends messages via mRNA to the cytoplasm to construct proteins explains the relationship between DNA and proteins. DNA to mRNA to tRNA and assembly of proteins is a one-way street, so important that it is called the Central Dogma of inheritance. Information does not come directly from the environment to affect genes. Genes change via mutations and affect the structure of individuals and how they interact with the environment. In the next lab, we show how DNA replication affects, how new cells are formed, and how genetic information is inherited by offspring.

Laboratory Report Sheet

NAME______________________________________ **SECTION** ______________ **GRADE** ______________

DNA Structure and Function

Objective

Explain the connections between DNA structure and function. Describe the processes of transcription and translation. Explain the importance of transcriptional regulation.

1. A nucleotide is the basic structure of the DNA molecule. Identify the structures below. Label the following nucleotide diagram using terms "sugar," "phosphate," and "base."

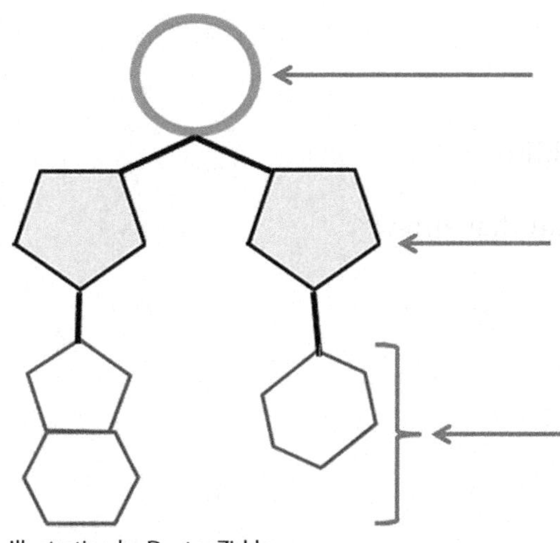

Illustration by Dexter Zirkle

2. Which molecule (DNA or RNA?) is single stranded? Which molecule is double stranded?

   Single strand =

   Double strand =

3. DNA utilizes these four nucleobases __________, __________, __________, __________

   RNA utilizes these four nucleobases __________, __________, __________, __________

4. In DNA, what base does adenine always connect to?____________________________

   In RNA, what base does adenine always connect to?____________________________

   In DNA, what base does guanine always connect to?____________________________

   In RNA, what base does guanine always connect to?____________________________

5. Who discovered DNA in 1869?
   a. Francis Crick
   b. Marie Curie
   c. Rosalind Franklin
   d. Friedrich Miescher
   e. Dmitri Mendeleev

6. Who discovered the *structure* of DNA in 1953?
   a. Friedrich Miescher
   b. James Watson
   c. Francis Crick
   d. Both Freidrich Miescher and James Watson
   e. James Watson, Francis Crick, and Rosalind Franklin

7. Which of the following items contain DNA? (circle all that apply)
   a. Celery
   b. Mushrooms
   c. Chimpanzee
   d. Bacteria
   e. Coral
   f. Jellyfish
   g. Many viruses

8. Match the DNA bases by entering the correct base in the nucleotides on the bottom half of the figure:

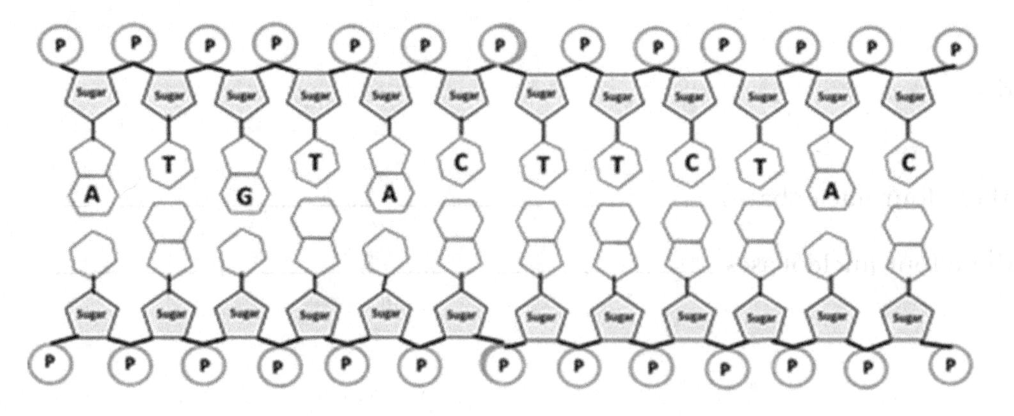

DNA Replication

9. How does the *double-stranded structure* of the DNA molecule contribute to making nearly identical copies (study Figures 8A.5 and 8A.6)?

10. **DNA Replication.** Using the DNA replication diagram below, fill in the appropriate bases (in the open circles) which correspond to the Leading Strand. Also, fill in the appropriate bases on the Lagging Strand as well as the bases which correspond to them (in the open circles).

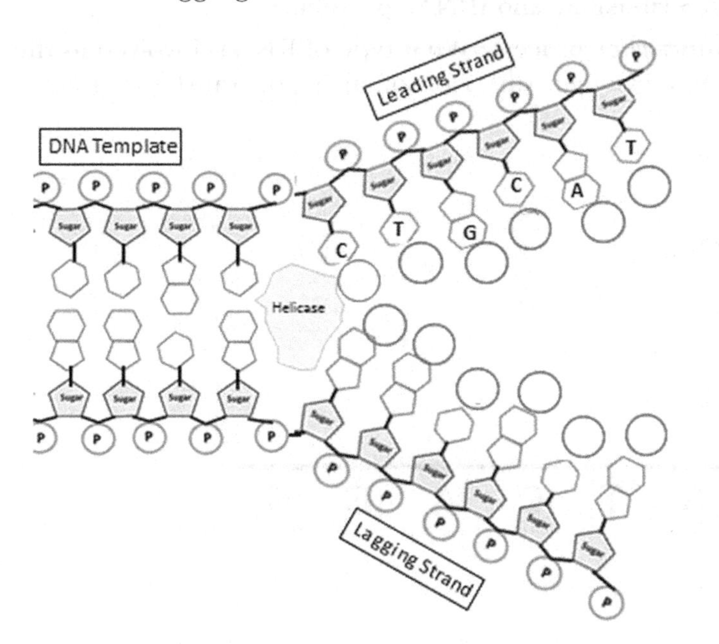

11. Given the following Template Strand (below), draw the two resulting strands of replication in the space to the right of the figure below.

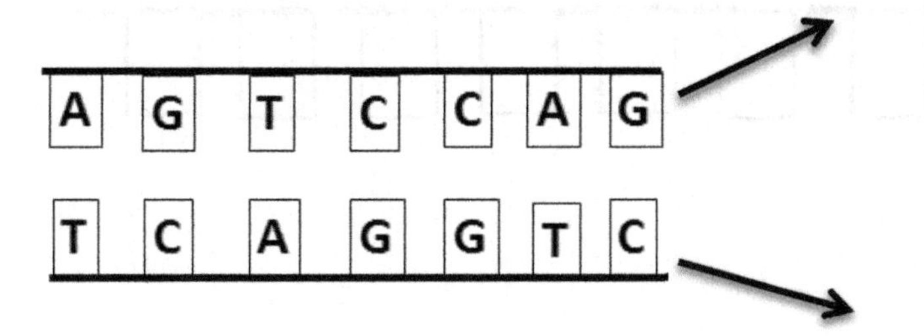

12. What are the three main differences between DNA and RNA?

    a.

    b.

    c.

13. **Discussion question:** Each of us began life as a single cell which divided into all of the cells in the adult body. Each cell is specialized as mature cells, and undergoes DNA replication before cell division. Is the DNA in your liver cells the same as the DNA in your skin cells?

Transcription and Translation

Transcription and translation: DNA > mRNA > ribosome and tRNA> protein.

14. What type of RNA is involved in the *transcription* process? What type of RNA is involved in the *translation* process? (Hint: choose among messenger (m), ribosomal (r), or transfer (t) RNA.)

    Transcription = _____RNA

    Translation = _____RNA & _____RNA

15. Given the following *DNA* sequence

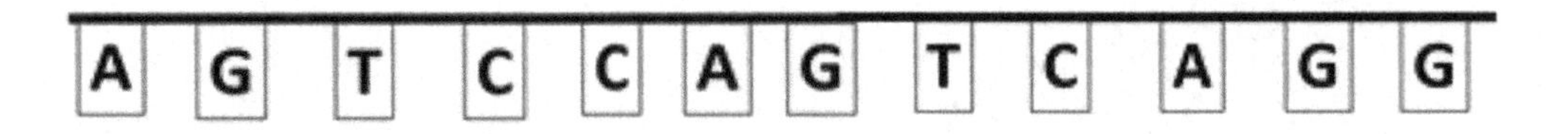

If the *section of the gene* in question 15 underwent transcription to an mRNA molecule, what would the *mRNA* sequence be?

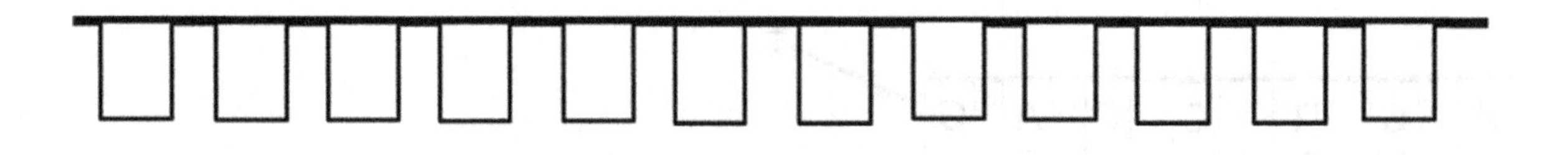

16. What *mRNA bases* would correspond to the DNA Template bases listed? Fill in the bases on the strand indicated by the arrow.

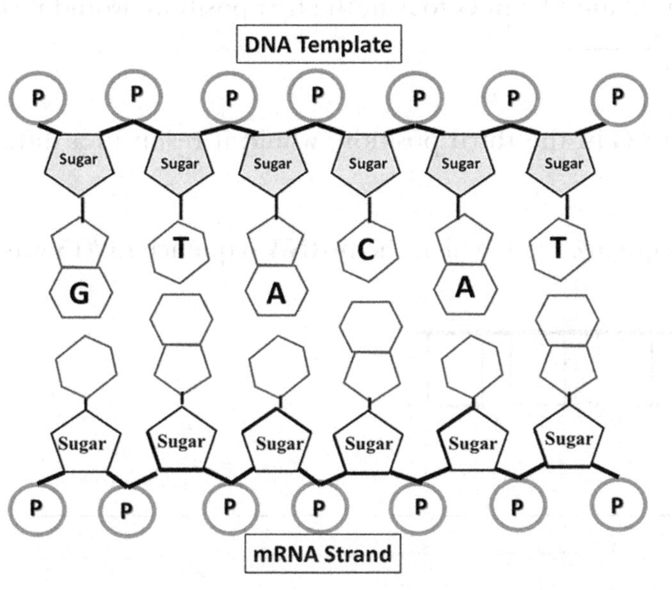

17. Given the following sequence of mRNA, identify the individual codons (reading left to right).

Using these mRNA codons (above), what is the *tRNA anticodon* sequence?

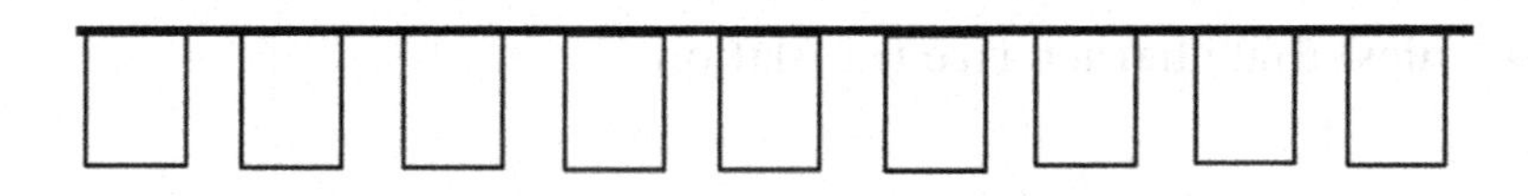

18. Which amino acids are being carried by each tRNA molecule in question 17? (use Figure 8A.9 to determine the amino acids involved).

________________, ________________, ________________

19. Given the following mRNA excerpt, and using Fig 8A.9, what are the amino acids represented in the mRNA?

UCA GGU CAG UCC

________________, ________________, ________________

20. Is there more than one CODON that will code for a specific amino acid?________________

21. What amino acid would be coded for given the codon **GUA**?________________

Given the codon **GUA**, if there were a mutation from **G** to **A** in the first position, would it result in a different amino acid?________________

If you introduced a mutation from **A** to **C** in the third position, would it result in a different amino acid?________________

22. What was the original **DNA** template sequence from which the **mRNA** sequence **in #15** was constructed?

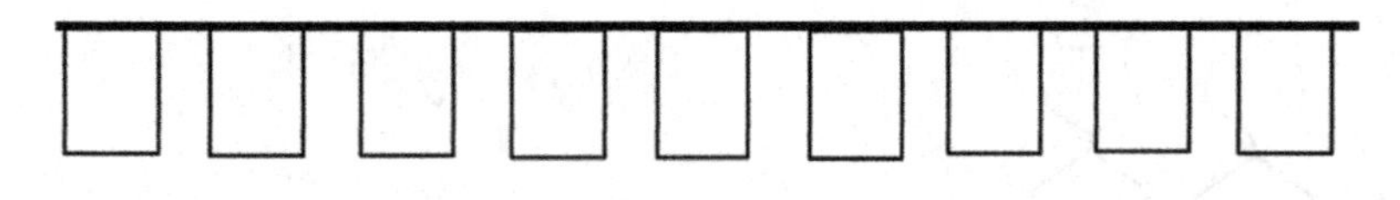

23. What is a gene? ________________________________

____

____

____

24. Reflection point! Identify two "rules" that characterize translation.

a.

b.

25. Does *transcription* occur inside or outside of the nucleus?

26. Does *transcription* utilize *tRNA*? (Y/N)

27. Does *translation* occur inside or outside of the nucleus?

28. Does *translation* utilize **tRNA**? (Y/N)

29. The final products of *translation* are proteins. (T/F)

30. Proteins are made up of amino acids. (T/F)

Evo-Devo (Evolutionary Development)

Mutations in the transcription factor FOXP2 have been linked with speech and language disorders in humans, but the gene is ancient and would have had other (possibly similar) functions in other animals.

31. Compare the DNA sequences below. Circle bases that are different among the codons. A macaque is a monkey that lives in Asia and originated in Africa.

Mouse: ATG CAA TAC GCC GCC GAA TCG TAC TGC ACG CAA GAA GCA GAC

Macaque: ATG CAT TAC GCC GCC GAA TCG TAC TGC ACG CAA GAA GCA GAC

Chimp: ATG CAT TAC GCC GCC GAA TCG TAC TGC ACG CAA GAA GCA GAC

Human: ATG CAT TAC GCC GCC GAA TCG TAC TGC ACG CAA GAC GCA GAG

32. How many amino acids are different between mouse and macaque? ____________________

33. How many codons are different between mice and humans?____________________

Between chimpanzees and humans?____________________

34. Which two organisms are most closely related and which two are most distantly related? What evidence supports your predictions?

35. What else would you like to know to strengthen your conclusions in #34?

Individuals of different species can express differences even if the structural gene is identical. Regulatory genes act as on/off switches for structural genes, regulate the rate and timing of transcription. In this example, FOXP2 is thought to regulate the transcription of CNTNAP2, a neural gene involved in speech development.

36. The gene sequence of CNTNAP2 is identical between chimpanzees and human. What does this suggest about the importance of FOXP2, and transcription factors in general, in achieving evolutionary differences?

37. It is often said that chimpanzees and humans share 99.8% of their genome. Given this minor difference, you would think that chimps and humans were nearly identical. They are not, of course, so wherein lies the true difference?

DNA Extraction and Electrophoresis

INTRODUCTION

We are going to explore DNA in two ways today through extraction from a banana and comparing DNA from samples collected from two different people.

ACTIVITY 1: DNA Electrophoresis[1]

In this exercise, you will simulate the forensic analysis of DNA evidence from a crime. The process begins by collecting human cell evidence from a crime scene and locating and collecting cell samples from suspects. Once the forensic scientist receives the various cell samples, his or her first task is to isolate the DNA from each of the samples. The scientist then needs to compare the portions of each DNA sample that are most likely to be unique to each person. This is usually accomplished by amplifying (increasing) copies of the critical portions of the DNA by using a technique known as the polymerase chain reaction (PCR).

Once amplified, the DNA samples are loaded into an agarose gel, which is then electrophoresed: A power supply provides an electrical field across the gel that "pulls" the DNA fragments away from the gel's loading wells toward a positive electrode. Smaller DNA fragments move faster than larger ones, so the different sizes will separate into bands during electrophoresis. This pattern of bands is referred to as the *DNA fingerprint*. Because each person's DNA produces a unique set of fragments, the suspects' DNA fingerprints can now be compared to the DNA fingerprint from the crime scene evidence to see if there is a match.

What Is PCR?

Most DNA samples found at crime scenes have too small an amount of DNA present for analysis. The PCR process amplifies a segment of DNA that varies from person to person and that lies between two regions of a known and stable nucleotide sequence. The process supplies the necessary primers, DNA polymerase, and nucleotide building blocks for the targeted replication process to occur.

The DNA molecule is denatured into two single strands by heating, then cooled to anneal (attach) the primers to the target sequence of DNA. The primers are short, chemically synthesized pieces of DNA that will act as start and stop signals for the DNA polymerase enzyme. The DNA polymerase then catalyzes the attachment of new nucleotides to each single strand, resulting in two double-stranded DNA molecules.

[1]Adapted from: "PCR Forensics Simulation Kit 21–1210," Carolina Biological Supply Company, © 1994.

The cycle of denaturation by heating, annealing the primers, and DNA synthesis is repeated numerous times to achieve an exponential growth in the amount of desired DNA product: One DNA segment replicates to form two segments, two to four, four to eight, eight to sixteen, etc. (see Figure 8B.1).

PCR first was developed in 1985 by Kary Mullis, and it has revolutionized our ability to analyze DNA evidence. Today, the process is an automated one, thanks to the use of a heat-stable form of DNA polymerase, discovered in a hot springs bacterium named *Thermus aquaticus*. This thermostable ability means that the DNA polymerase can be repeatedly heated and cooled without destroying its activity.

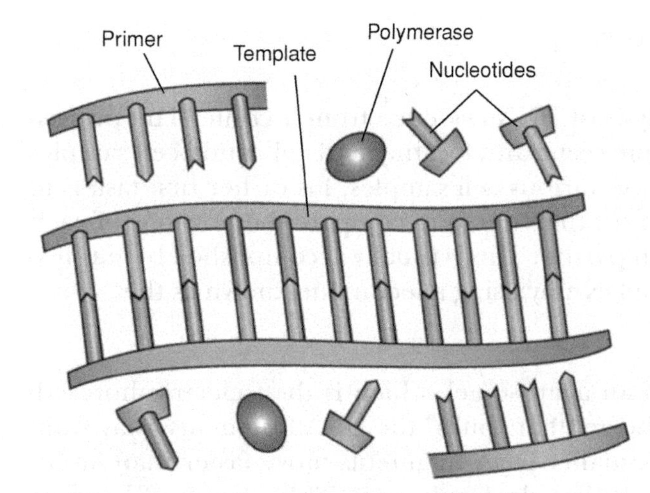

A. Test tube containg DNA strand fragments (templates), complimentary fragments (primers), single nucleotides and polymerases.

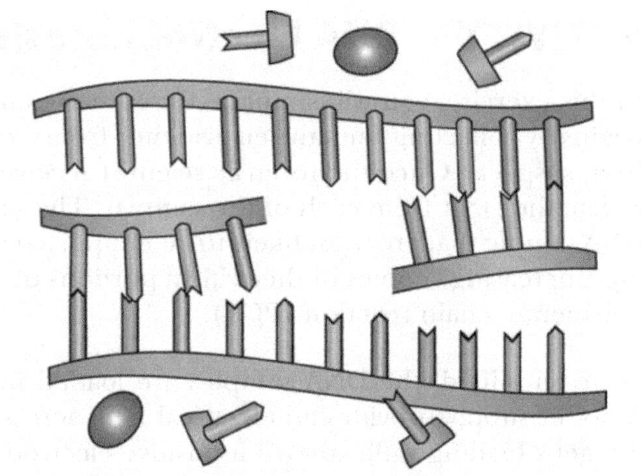

B. Solution heated to 95 C, causing DNA strands to separate. Solution is then cooled to 37 C, and primers attach to complimentary sequences on each template strand.

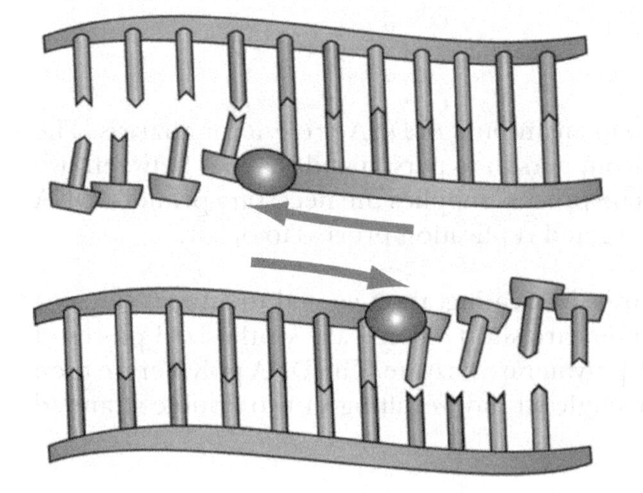

C. Solution reheated to 72 C, causing polymerases to attach to primer ends and create new DNA strands using single nucleotides.

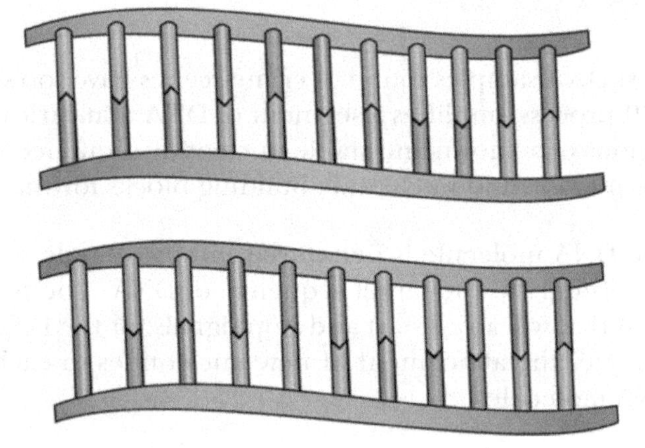

D. Two identical copies of original DNA fragment. Several more cycles follow, doubling number of DNA fragments each time.

Figure 8B.1

What Portion of the DNA Molecule Is Used for PCR?

While we tend to focus our attention on the portion of the DNA that functions as genes, much of eukaryotic DNA is non-coding. This non-coding portion of the DNA varies in nucleotide sequence much more than the genetic portion does. Why?

One common type of variation occurs when a short sequence of nucleotides repeats itself over and over (see Figure 8B.1). Although everyone has the same types of repeated sequences at the same places in their chromosomes, the number of repeats varies tremendously. Referred to as variable number of tandem repeats (VNTRs), the function of these repeated sequences is not understood, but several have been well studied.

> Example of short nucleotide sequence: 5'-GAAGG-3'
>
> How many tandem repeats occur in this strand?:
> 5'– GAAGGGAAGGGAAGGGAAGGGAAGG –3'

Figure 8B.2 Tandem Repeats

If we use PCR to amplify a certain VNTR from several DNA samples, then separate the products using electrophoresis, we can study the pattern formed from the VNTRs of each sample. By selecting many different positions of VNTRs on a genome, we can see a unique pattern, or fingerprint, for each individual. In forensics, VNTR patterns obtained from DNA at crime scenes can be compared to the DNA of suspects to determine if there is a match.

Procedure

You will work in groups to analyze the DNA evidence from a simulated crime scene. The materials you will be using include purified DNA samples that have been prepared to simulate the products of PCR.

A brief scenario for our crime could be the following . . .

> A man was beaten to death at night when leaving an ATM. An eyewitness clearly saw a man fl eeing the scene and almost identified him from pictures of suspects, but it turned out the suspect had a brother he strongly resembled who was also pictured. The eyewitness couldn't distinguish the suspect from his brother, even though they weren't twins. Fortunately, the witness remembered seeing the victim claw and scratch the attacker, leaving cells containing DNA under the fingernails. Even though several months had passed before the brothers were found, the attacker could be identified by the DNA sample taken from the victim.

Procedure A: Casting Agarose Gel

1. Assemble gel casting tray: Attach dams on both ends of tray, place well-forming comb in the notch **closer** to the dams (not in the middle of tray). Put tray in a spot out of the way so that it can remain undisturbed once the agarose has been poured.
2. The agarose tubes have been pre-measured for the casting trays. They are in a 65°C water bath, which is about 5°C hotter than the ideal pouring temperature.
3. Pour the agarose into the assembled casting tray. The gel should cover approximately half the height of the comb teeth. You can use a dissecting probe to move any large bubbles or solid debris to the end of the tray (opposite comb) while the gel is still liquid.
4. Avoid moving the gel for about 20 minutes to allow it to solidify. The gel will become cloudy during this time.

5. When the gel has solidified, carefully remove the comb. You may need to wriggle it *gently* so that it slides out of the solid agarose. Next, run a probe along each dam to prevent the gel from sticking and ripping. Carefully remove the dams. Place the gel in the electrophoresis chamber (making sure the gel does not slide out of the tray) with the wells at the negative (black) end. The level of buffer in the chamber should just cover the gel surface.
6. You are now ready to load your DNA samples. A few practice gels have been set up with "practice loading dye" to familiarize you with pipetting and the gel-loading process.

Procedure B : Gel Loading

1. Each gel will need one strip of the six DNA samples, a micropipet (50 µL size), and micropipet tips.
2. Tap the strip gently on the bench to get all the contents at the bottom. Using new tips for each sample vial, load the wells in this order:
   a. DNA from crime scene cut with Enzyme 1
   b. DNA from crime scene cut with Enzyme 2
   c. DNA from Suspect 1 cut with Enzyme 1
   d. DNA from Suspect 1 cut with Enzyme 2
   e. DNA from Suspect 2 cut with Enzyme 1
   f. DNA from Suspect 2 cut with Enzyme 2

Load the entire tube contents in the well; be careful not to expel any air bubbles while pipetting. The samples have loading dye in them already that is heavier than the buffer, so the sample will sink to the bottom of the well. Be careful **not to puncture** the tip through the bottom of the gel!

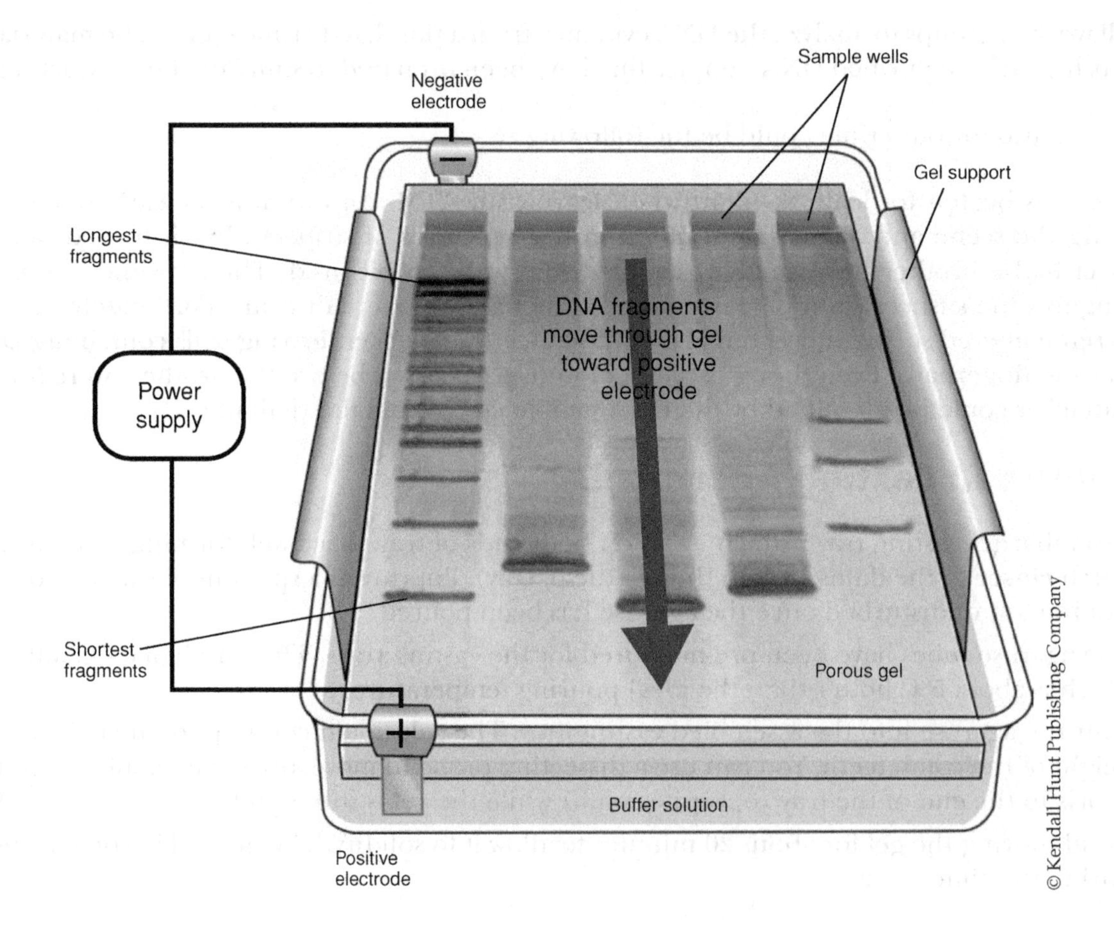

Procedure C : Electrophoresis

1. When done loading all samples, place the chamber cover on, matching up red to red and black to black. Attach the electrodes to the same channel of the power supply.
2. Turn the power supply on. The voltage should be set at 125 volts. When the current is successfully going through the chamber, you will be able to see tiny bubbles rising from the chamber wires.
3. Allow the gel to run for approximately 30 minutes. The blue tracking dye will run ahead of the DNA samples; you can stop when it gets toward the end of the gel.
4. Turn **off** the power supply, remove the electrodes from the power supply and lift the cover off the chamber. Carefully remove the gel from the chamber and place it in a staining tray for observation.

Procedure D : Visualizing Bands and Staining

1. Pour approximately 100 to 125 mL of 1X FlashBlue stain. Stain for only **4–5 minutes.** Carefully pour the stain back into its container to reuse.
2. You will need to destain the gel for better visualization of bands. Place gel (or use the staining tray) in a rinsing pan. Fill with tap water. Destaining should take only 10 minutes, but you may want to gently agitate the pan and change the water if it is still very blue. If after 10 minutes, visualization is still difficult, you can repeat the destaining process (it can't be over-destained).
3. Place gel on a light tray and examine the bands of DNA in the gel and compare the DNA from the crime scene and the two suspects for both Enzyme 1 and Enzyme 2.

ACTIVITY 2: Extracting DNA from Bananas[2]

DNA can be extracted from the nucleus of cells by adding an aqueous buffered extraction solution to tissue. The cells are chemically lysed (or broken open) and the DNA is released. DNA is soluble in water but not in alcohol. Since isopropyl alcohol (AKA rubbing alcohol, isopropanol) has a lower density than water, it will form a layer above the DNA solution. A glass rod is used to spool the two liquids at their interface to separate the DNA from the solution. The DNA will appear as a viscous mass on the spooling rod.

Procedure

1. Carefully slice a piece of banana about 1 cm^2 and place it in a mortar.

2. Add 5 ml of DNA extraction buffer to the mortar and use the pestle to mash the banana tissue and buffer together. This releases the cellular contents and DNA from the cells.

[2] *Adapted from: Edvotek, Inc.,* "Do Onions, Strawberries, and Bananas have DNA?"EDVO-kit# S-75

3. Obtain a self-standing test tube and place a funnel in the top. Now pour the solution from the mortar through a sieve into the funnel and test tube. This will filter the large pieces of banana tissue from the extraction tissue.

4. Overlay the solution with 5 ml of **cold** isopropyl alcohol, by pouring slowly down the side of the tube. Do not mix the two solutions but allow them to layer themselves. (The isopropyl alcohol should be kept in the fridge when the class is done using it.)

5. Place a glass rod into the test tube and twirl it gently at the interface of the two layers. The DNA will begin to spool around the glass rod.

6. After spooling for several minutes, remove the rod to observe the extracted DNA.

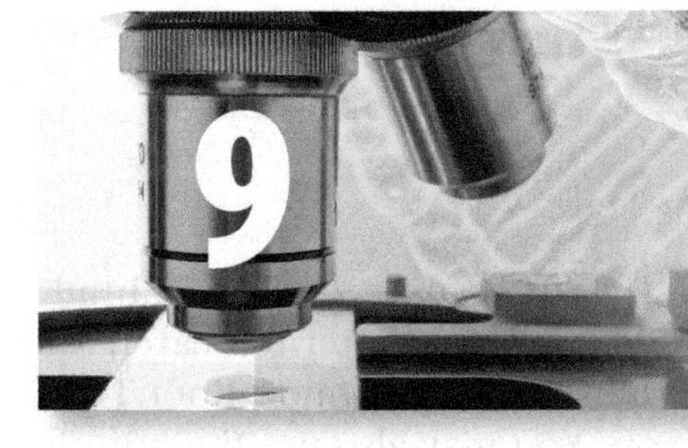

9 Virtual Laboratory—Mitosis and Meiosis

Introduction

You may work together with your lab partner, but your written responses to the questions and your drawings must be your own. Otherwise, you will receive no credit for this lab. Make sure you understand all aspects of this lab. There will be questions from this lab on your quizzes and tests.

Overview

In this virtual laboratory, you will use some resources available on the Internet to investigate the process of mitosis and meiosis. The first part is a study of mitosis. You will examine slides of whitefish and onion cells undergoing mitosis. You will be able to make comparisons of mitosis in plant and animal cells. The second part is a study of meiosis. You will examine slides of meiosis taking place in *Lillium spp.* anthers. These plants are related to the familiar traditional Easter lilies. Finally, you will make graphic comparisons between cells undergoing mitosis and those undergoing meiosis. You will be taking some online self-quizzes to test your knowledge of both meiosis and mitosis.

Procedures and Analysis Questions

A. Mitosis in Onion Root-Tip Cells and in Whitefish Cells

Go to the website: http://biog-1101-1104.bio.cornell.edu/biog101_104/tutorials/cell_division.html. To find the site easily without having to type in the long web-page address, do an Internet search for "Cornell Biology Mitosis," and select "Cell Division Tutorials" and then "Review—Onion Root Tip." If this link is no longer active, the instructor will provide one for you. You may also use the mitosis images in your textbook.

1. Sketch the slides that you see at this website for onion root-tip cells undergoing mitosis on the laboratory report form. (See: Review—Onion Root Tip.) Label all the parts of yoursketch.

Laboratory Report Sheet

NAME________________________________ **SECTION** ____________ **GRADE** ____________

Virtual Laboratory—Mitosis and Meiosis

Onion Root-Tip Mitosis

The nondividing cell is in a stage called **interphase**. The nucleus may have one or more dark-stained **nucleoli** and is filled with a fine network of threads, the **chromatin**. During **interphase**, DNA replication occurs.

INTERPHASE DRAWING

The first sign of division occurs in **prophase**. There is a thickening of the chromatin threads, which continues until it is evident that the chromatin has condensed into **chromosomes**. With higher magnification you may be able to see that each chromosome is composed of two **chromatids** joined at a **centromere**. As prophase continues, the chromatids continue to shorten and thicken. In late prophase the nuclear envelope and nucleoli are no longer visible, and the chromosomes are free in the cytoplasm. Just before this time the first sign of a spindle appears in the cytoplasm; the spindle apparatus is made up of **microtubules,** and it is thought that these microtubules may pull the chromosomes toward the poles of the cell where the two daughter nuclei will eventually form.

PROPHASE DRAWING

At **metaphase**, the chromosomes have moved to the center of the spindle. One particular portion of each chromosome, the centromere, attaches to the spindle. The centromeres of all the chromosomes lie at about the same level of the spindle, on a plane called the metaphase plate. At metaphase you should be able to observe the two chromatids of some of the chromosomes.

METAPHASE DRAWING

At the beginning of **anaphase**, the centromere regions of each pair of chromatids separate and are moved by the spindle fibers toward opposite poles of the spindle, dragging the rest of the chromatid behind them. NOTE THAT once the two chromatids separate, each is called a **chromosome**. These daughter chromosomes continue poleward movement until they form two compact clumps, one at each spindlepole.

ANAPHASE DRAWING

Telophase, the last stage of mitosis, is marked by a pronounced condensation of chromosomes, followed by the formation of a new nuclear envelope around each group of chromosomes. The chromosomes gradually uncoil to form the fine chromatin network seen in interphase, and the nucleoli and nuclear envelope reappear. **Cytokinesis** may occur. This is the division of the cytoplasm into two cells. In plants, a new cell wall is laid down between the daughter cells. In animal cells, the old cell will pinch off in the middle along a **cleavage furrow** to form two new daughter cells.

TELOPHASE DRAWING

2. Sketch the slides that you see at this website or your textbook for whitefish cells undergoing mitosis. (See: Review—WhitefishMitosis.) The white fish blastula is excellent for the study of cell division. As soon as the egg is fertilized, it begins to divide, and nuclear division after nuclear division follows. Label all the parts of your sketch.

Interphase Drawing **Prophase Drawing**

Metaphase Drawing **Anaphase Drawing**

Telophase Drawing

3. Why is it more accurate to call mitosis "nuclear replication" rather than "cellular division"?

4. Explain why you think the whitefish (blastula) and onion root tip are selected for a study of mitosis?

5. What were the differences and similarities in the slides of onion root tip (plant) and whitefish blastula (animal) undergoing mitosis?

B. Time Involved in Cell Replication

To answer the questions for this section, go to the following website: http://www.biology.arizona.edu/cell_bio/activities/cell_cycle/assignment.html. Note: If the website is not working, your instructor will provide an alternate.

1. What is the hypothesis (or prediction) of this exercise?

2. Complete the following data table based on the data presented at the website.

Stage of Mitosis	Number of Cells Counted	Percent (of total number of cells)	Minutes of cell cycle spent in this stage
Interphase			
Prophase			
Metaphase			
Anaphase			
Telophase			

On average, it takes 24 hours (or 1,440 minutes) for onion root-tip cells to complete the cell cycle. You can calculate the amount of time spent in each phase of the cell cycle from the percent of cells in that stage in the following way:

Percent of cells in that stage/100 * 1,440 minutes = ________ minutes of cell cycle spent in that stage.

3. If your observations had not been restricted to the area of the root tip that is actively dividing, how would your results have been different?

4. Based on the data in the above table, what can you infer about the relative length of time an onion root-tip cell spends in each stage of cell division?

C. Meiosis

In this part of the lab, you will examine slides of *Lillium spp.* undergoing meiosis. Draw and label what you see in the slide for each stage of meiosis. Provide a listing of the main activities that are occurring in the cell in each stage, as indicated on the following website: http://www.dmacc.cc.ia.us/instructors/lillium.htmor from your textbook.

MEIOSIS I DRAWING

Prophase I Drawing

Metaphase I Drawing

Anaphase I Drawing

Telophase I/Cytokinesis Drawing

MEIOSIS II DRAWING

Prophase II Drawing

Metaphase II Drawing

Anaphase II Drawing

Telophase II/Cytokinesis Drawing

D. Gamete Development

1. Explain how eggs develop from meiosis. What is meant by a polar body? Why is it beneficial that the eggs that are formed are relatively large? How is this process different from spermatogenesis?

E. Comparison of Meiosis and Mitosis

Examine the following website for an animated comparison of mitosis and meiosis: http://biology.clc.uc.edu/courses/bio104/meiomito.htm. Alternatively, use your textbook to answer the following questions.

1. What is the purpose or goal of mitosis versus meiosis? How are these two processes similar, and how are they different? List three major differences between the events of mitosis and meiosis.

2. Compare mitosis and meiosis with respect to each of the following:

	Mitosis	Meiosis
Chromosome number of parent cells		
Number of DNA replications		
Number of divisions		
Number of daughter cells produced		
Chromosome number of daughter cells		
Purpose		

F. Self-Test Exercises

Complete the test review questions at the following websites. Complete these questions until you fully understand them. Also, practice the questions in your textbook and the textbook's website.

1. Try to answer the review exercises for mitosis and meiosis from the following website. Make sure you see the pictures of the chromosome beads:

   http://www.und.nodak.edu/dept/jcarmich/101lab/lab9/mittest.html

2. http://biog-101-104.bio.cornell.edu/BioG101_104/tutorials/cell_division.html
   (This website includes a total of 19 questions you should try.)

3. http://www.biology.arizona.edu/cell_bio/tutorials/meiosis/problems.html
   (This website includes a total of 10 questions you should try.)

10 Classification and Survey of the Kingdoms: Prokaryote, Protista, and Fungi

Objectives

To be able to

1. Define taxonomy.
2. Describe the binomial system of nomenclature and define species.
3. State the major categories of classification from the largest to the smallest.
4. List the five kingdoms and the major identifying characteristics according to possession of a true nucleus, structural organization, type of nutrition, method of reproduction, and mode of locomotion.
5. Recognize the two major groups belonging to the Prokaryote as bacteria and cyanobacteria and describe their main characteristics.
6. Identify the three shapes of bacteria.
7. State the function of heterocyst.
8. Differentiate among the groups belonging to the Protista: animal-like protists, plant-like protists; and fungal-like protists.
9. List three methods of locomotion of the animal-like protists observed and examples of each.
10. Differentiate among three phyla of the plant-like protists: Bacillariophyta, Chlorophyta, and Phaeophyta and recognize examples of each.
11. Describe the structure of a fungal-like protist and recognize an example.
12. Describe the main characteristics of the Fungi.
13. Differentiate among three phyla of fungi: Zygomycetes, Ascomycetes, and Basidiomycetes and recognize examples of each.
14. Define: hyphae, mycelium, zygospore, sporangium, ascospores, asci, conidiophores, basidiospores, cap, stipe, and gills.
15. Recognize all the organisms observed.
16. Complete the Laboratory Report Sheet.

Introduction

In order to learn massive amounts of information, man attempts to organize it. To identify the great diversity of biological organisms, a system of classification has been devised. The study of classification is known as taxonomy and scientists that specialize in the field of classification are called taxonomists.

The process of classifying organisms entails placing organisms together in groups. Many types of characteristics may be used, such as appearance, behavior, pigments, size, habitat, etc. In the mid-1700s, a Swedish botanist, Carolus Linnaeus, developed a system of classification. Each organism is given two Latin names: one a generic name, the genus; the other a specific name or species. The generic name indicates a group of similar organisms and the specific name indicates a particular type of organism. This system of using both parts of the name to identify the organism is referred to as the binomial system of nomenclature. The genus is always placed first and capitalized, whereas the species is not. Both names are italicized or underlined. For example, there are many species of pine trees, all of the genus *Pinus;* the white pine is *Pinus strobus,* whereas the black pine is *Pinus nigra.*

The system of classification used also assigns species to a hierarchy of increasingly general groups, to finally the most general and largest group, the kingdom. In this hierarchial system, a species is a group of organisms that are of the same type, have the same characteristics, and can successfully interbreed. Closely related species belong to the same genus. Closely related genera belong to the same family; families are grouped into orders; orders into classes; classes into phyla; and phyla into kingdoms. Each of these levels of the classification hierarchy may be referred to as a taxon. Therefore, the family and class are examples of a taxon. Within each taxon there may be subclassifications such as subphylum and suborder. The hierarchy from largest to smallest may be remembered easily by recalling the following nonsensical sentence: **"King Phillip Came Over From Great Scotland."** The first letter of each word is a reminder of the sequence: **Kingdom, Phylum, Class, Order, Family, Genus, and Species.**

Below is the classification of humans, giving the name of each of the major categories or taxons.

Taxon	**Man**
Kingdom	Animal
Phylum	Chordata
Class	Mammalia
Order	Primata
Family	Hominidae
Genus	*Homo*
Species	*sapiens*

Many biologists classify living organisms into five kingdoms: Prokaryote (Monera), Protista, Fungi, Plantae, and Animalia. These kingdoms and their major identifying characteristics are shown in the following table.

THE FIVE KINGDOM SYSTEM

Kingdom	Major Identifying Characteristics
Prokaryote (Monera)	Lack a distinct nucleus and other membranous organelles; usually unicellular; mainly heterotrophic; reproduction is usually asexual; generally nonmotile.
Protista	Eukaryotic; unicellular or multicellular showing relatively little division of labor and no tissue formation; most modes of nutrition are represented; reproduction is both asexual and sexual; numerous means of motility.
Fungi	Eukaryotic; mainly multicellular; heterotrophic; reproduction through asexual or sexual formation of spores; nonmotile.
Plantae	Eukaryotic; multicellular showing differentiated tissue and organs; photosynthetic autotrophs; asexual and sexual reproduction; nonmotile.
Animalia	Eukaryotic; multicellular, most showing differentiated tissues, organs and systems; heterotrophic; reproduction generally sexual; most motile.

In this exercise, representatives from the three kingdoms, Prokaryote, Protista, and Fungi, will be examined. Representatives from the kingdoms Plantae and Animalia will be studied in later exercises.

ACTIVITY 1: Kingdom: Prokaryote (Monera)

Examples of organisms belonging to the Prokaryote include bacteria and cyanobacteria (formerly blue-green algae). Bacteria are some of the smallest living organisms, which makes it difficult to see any detailed structure under the light microscope. However, they can be classified according to shape, a characteristic that is more easily identifiable. This classification includes: coccus (spherical); bacillus (rod-shaped); and spirillum (spiral-shaped). They may exist singly or in colonies as shown in Figure 10.1.

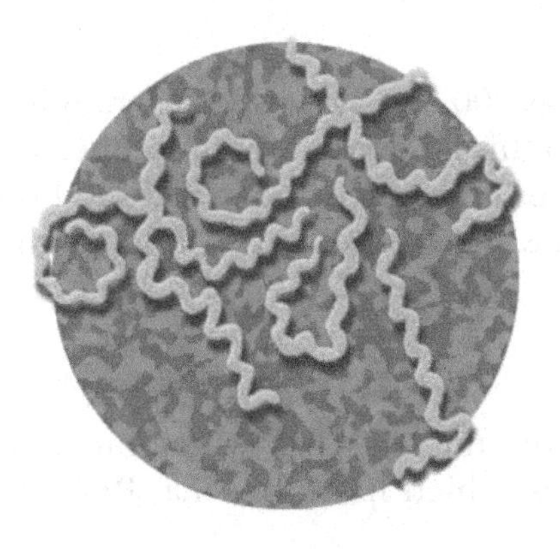

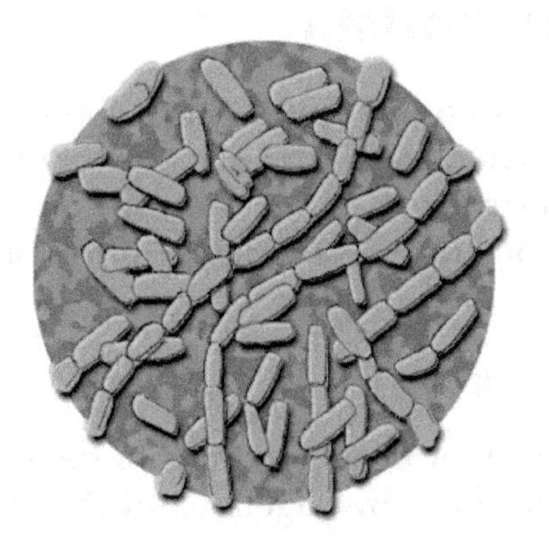

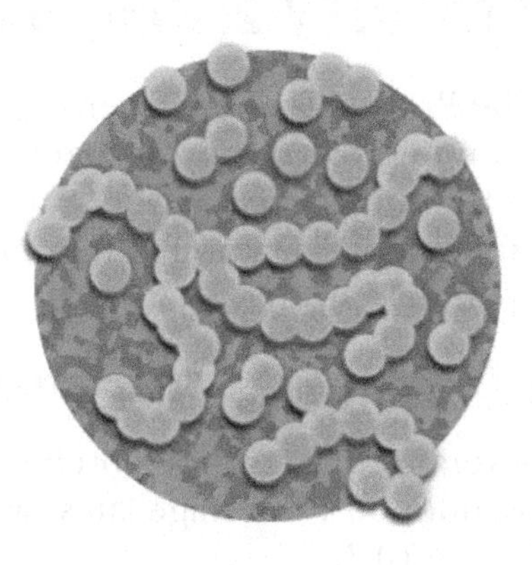

FIGURE 10.1 Bacterial Forms—© Kendall Hunt Publishing Company

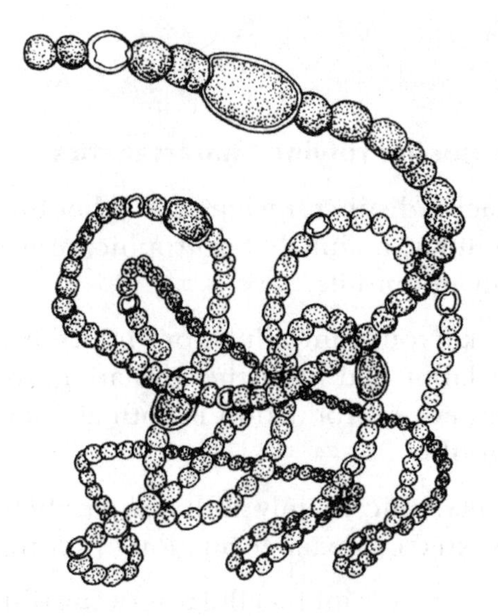

FIGURE 10.2 Anabaena—© Kendall Hunt Publishing Company

The cyanobacteria exist mainly as colonies or filaments and are aquatic or terrestrial. Chlorophyll is present and the organisms are photosynthetic. In addition, some of these are able to fix nitrogen which occurs in enlarged cells called heterocysts. Refer to Figure 10.2.

Procedure

1. Obtain a microscope slide of Bacteria, mixed culture. Examine the slide under low power, high power, and oil immersion. The cells are stained, but note the lack of a true nucleus. Note the different shapes of the bacteria: **coccus, bacillus,** and **spirillum** and diagram these in the space provided on the Laboratory Report Sheet.
2. Obtain a microscope slide of *Anabaena,* an example of a filamentous cyanobacteria. Examine the slide under low and high power. Note the organization of the cells and the slightly enlarged cells, the **heterocysts.** Diagram this organism in the space provided on the Laboratory Report Sheet.

ACTIVITY 2: Kingdom: Protista

The Protista is a very diverse group of organisms including as many as 200,000 living species, ranging from microscopic protozoans to giant algae. The phyla belonging to this kingdom are too numerous to list in this exercise, but vary in structure, mode of nutrition, means of reproduction, and method of locomotion, among other characteristics. Only a few of the phyla will be examined to indicate the great diversity.

A. ANIMAL-LIKE PROTISTS

Several groups will be observed to show three methods of locomotion: the amoebas, which move by pseudopods; the flagellates, which move by flagella; and the ciliates, which move by cilia. Refer to Figure 10.3.

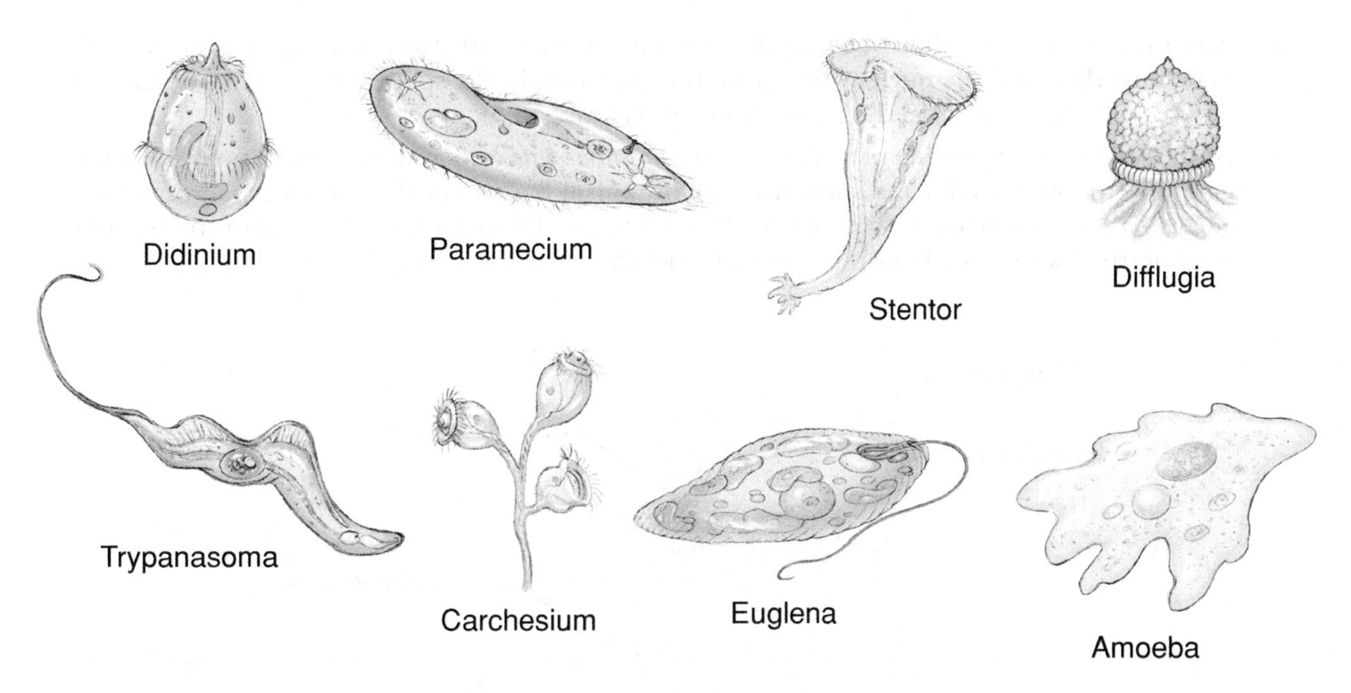

FIGURE 10.3 Protest Diversity—© Kendall Hunt Publishing Company

Procedure

1. Examine a microscope slide of *Amoeba* under low and high power. Note the lack of a definite shape and the presence of **pseudopods** for movement. The **nucleus** should be evident. Diagram the organism in the space provided on the Laboratory Report Sheet and label the parts observed. (Figure 10.4)

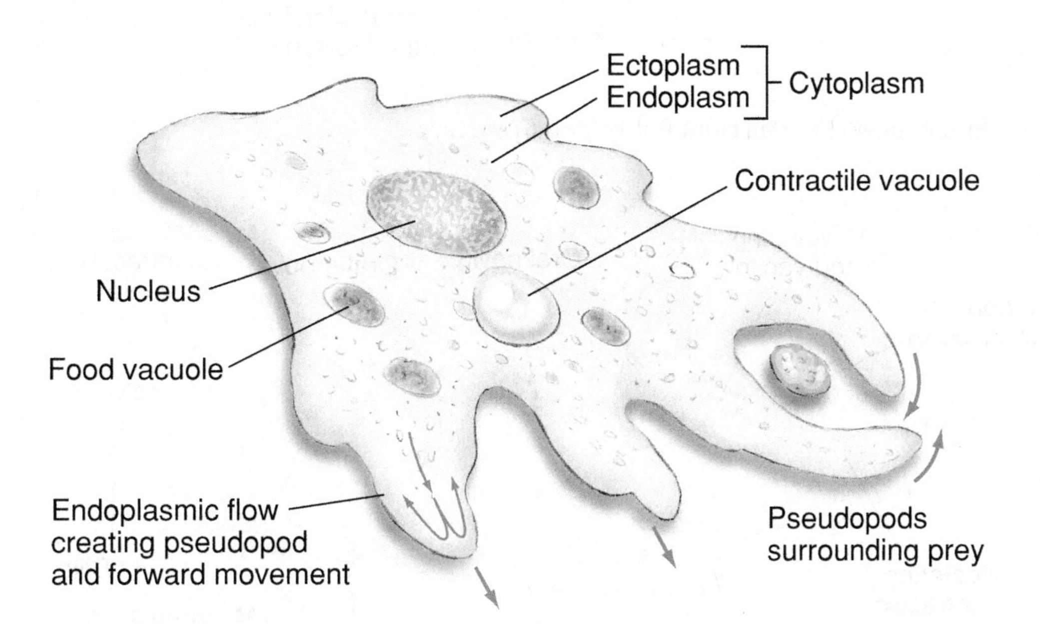

FIGURE 10.4 Amoeba—© Kendall Hunt Publishing Company

2. Examine a microscope slide of a flagellate such as *Euglena* under low and high power. Note the single **flagellum** and the **nucleus**. Diagram the organism in the space provided on the Laboratory Report Sheet and label the parts observed. (Figure 10.5)
3. Examine a microscope slide of a ciliate such as *Paramecium* under low and high power. Under high power, note the **cilia** and **vacuoles**. The organism has two nuclei, a larger macronucleus and a smaller micronucleus that may be difficult to see. Diagram the organism in the space provided on the Laboratory Report Sheet and label the parts observed. (Figure 10.6)

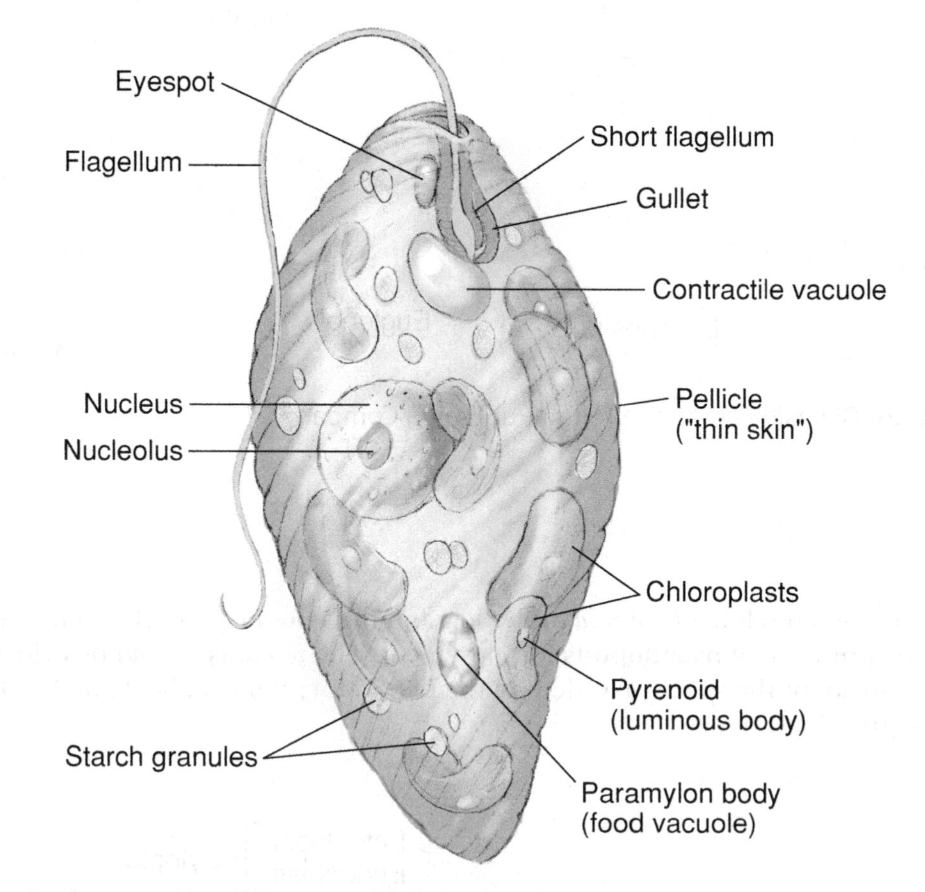

FIGURE 10.5 Euglena—© Kendall Hunt Publishing Company

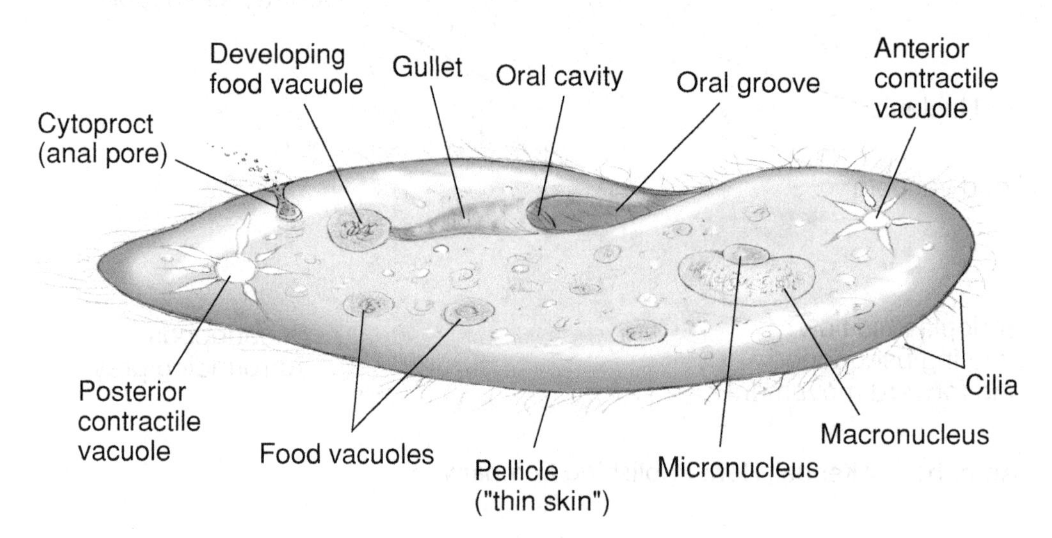

FIGURE 10.6 Paramecium—© Kendall Hunt Publishing Company

B. PLANT-LIKE PROTISTS

The plant-like protists or algae are capable of photosynthesis. To show their variety, examples of unicellular and multicellular organisms will be observed, sampling a few of the phyla. Examples of unicellular algae are diatoms (although some are colonial), comprising the phylum, Bacillariophyta. These organisms have a shell containing silica. Most are photosynthetic and are important producers in the aquatic ecosystem, making up an important part of the plankton.

The green algae, the members of the phylum Chlorophyta, include both unicellular, filamentous, and multicellular algae. This group also exhibits a range of reproductive methods, both asexual and sexual.

An example of a phylum of protists consisting of multicellular organisms is the Phaeophyta or brown algae. Some of the largest algae, the kelps, belong to this group. In addition to chlorophyll, they contain the brown pigment, fucoxanthin. The polysaccharide, algin, found in their cell walls, is used commercially as a thickening agent in various foods such as ice cream and puddings.

Procedure

1. Examine a slide of diatoms under low and high power. Note the shell and various shapes. Diagram a few examples in the space provided on the Laboratory Report Sheet. (Figure 10.7)
2. Examine a slide of *Spirogyra*. Observe the **filamentous body form** of these organisms, the **spiral-shaped chloroplast**, and the well-defined **nucleus** in each cell. Diagram the organism in the space provided on the Laboratory Report Sheet and label the parts observed. (Figure 10.8)
3. Observe a preserved specimen of the kelp, *Laminaria*. Observe the body structure consisting of a broad leaf **blade,** the **stem** or **stipe**, and the **holdfast** at its base that is used for attachment. (Figure 10.9)

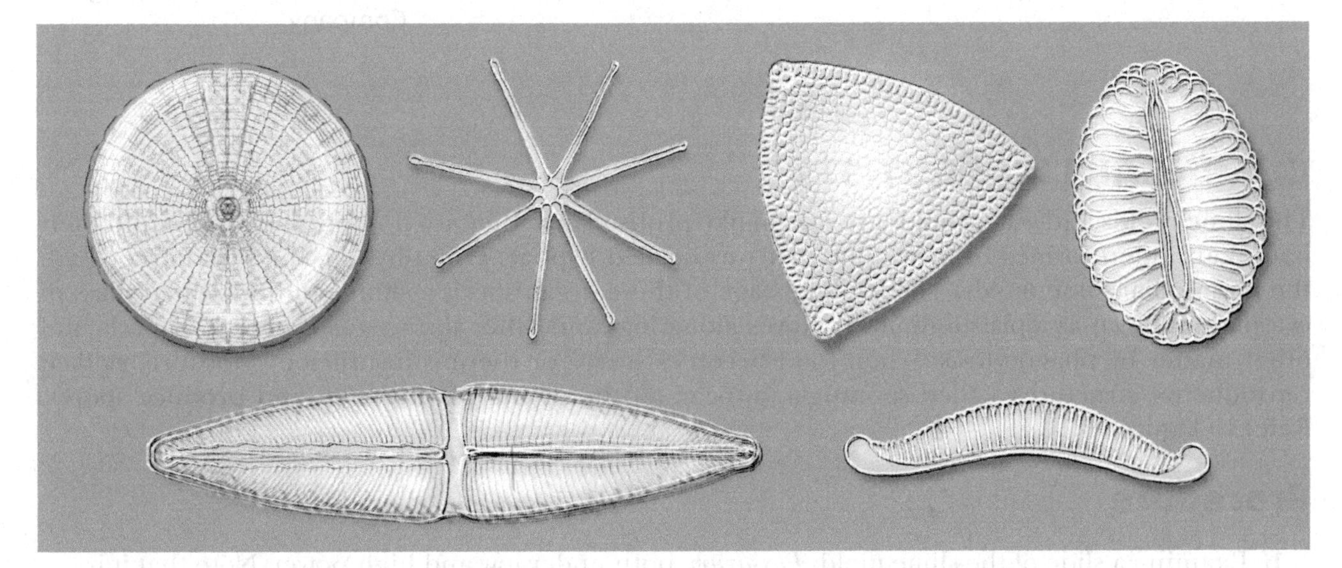

FIGURE 10.7 Diatoms—© Kendall Hunt Publishing Company

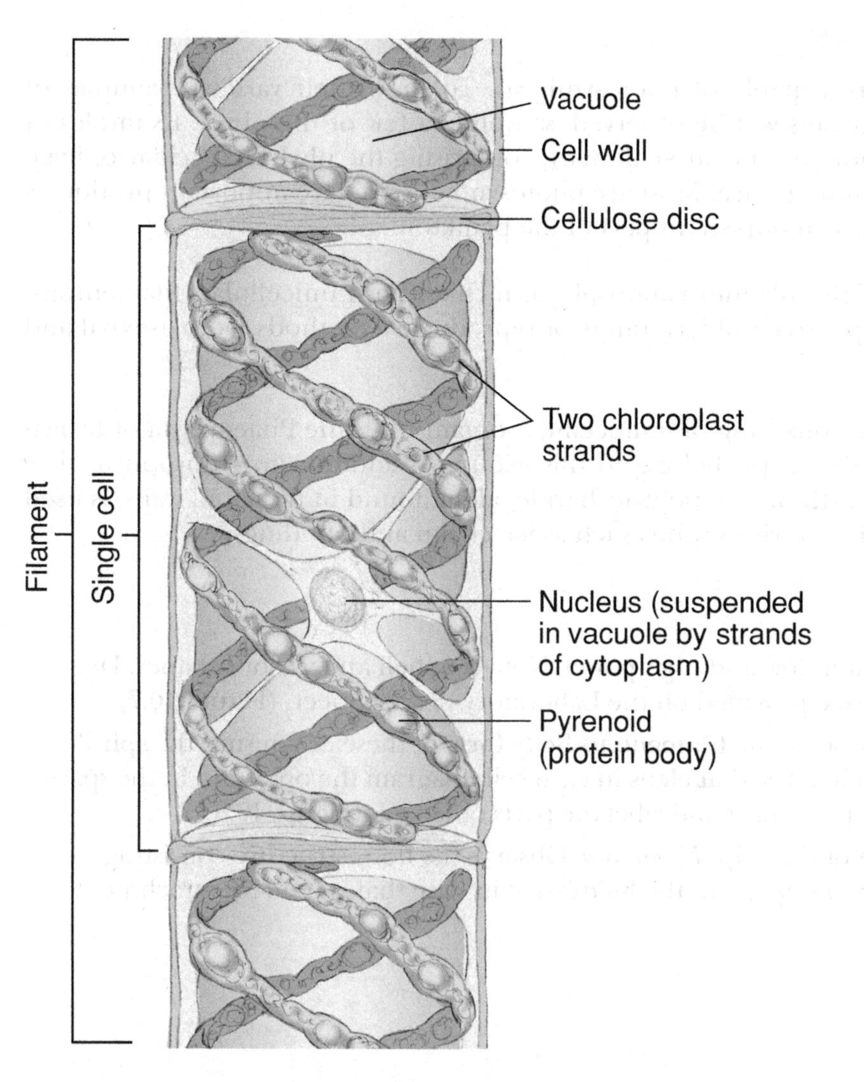

FIGURE 10.8 Spirogyra—© Kendall Hunt Publishing Company

FIGURE 10.9. Laminaria—© Kendall Hunt Publishing Company

C. FUNGAL-LIKE PROTISTS

The slime molds and water molds are fungal-like protists since they are non-photosynthetic and their bodies are often formed of threadlike structures called hyphae. One phylum, the Myxomycota, are the plasmodial slime molds. The feeding stage of these organisms is a slimy multinucleated mass of cytoplasm known as a plasmodium. It moves along logs and other surfaces as it ingests bacteria and other matter by phagocytosis. When food becomes scarce or there is insufficient moisture, stalked reproductive structures, called sporangia, appear on the drying plasmodium and produce spores. Refer to Figure 10.10.

Procedure

1. Examine a slide of the slime mold, *Physarum*, both under low and high power. Note that it is **multinucleated.** Observe its irregular mass of cytoplasm and the presence of vacuoles. Examine for the presence of **sporangia**. Diagram the organism in the space provided on the Laboratory Report Sheet and label the parts observed.

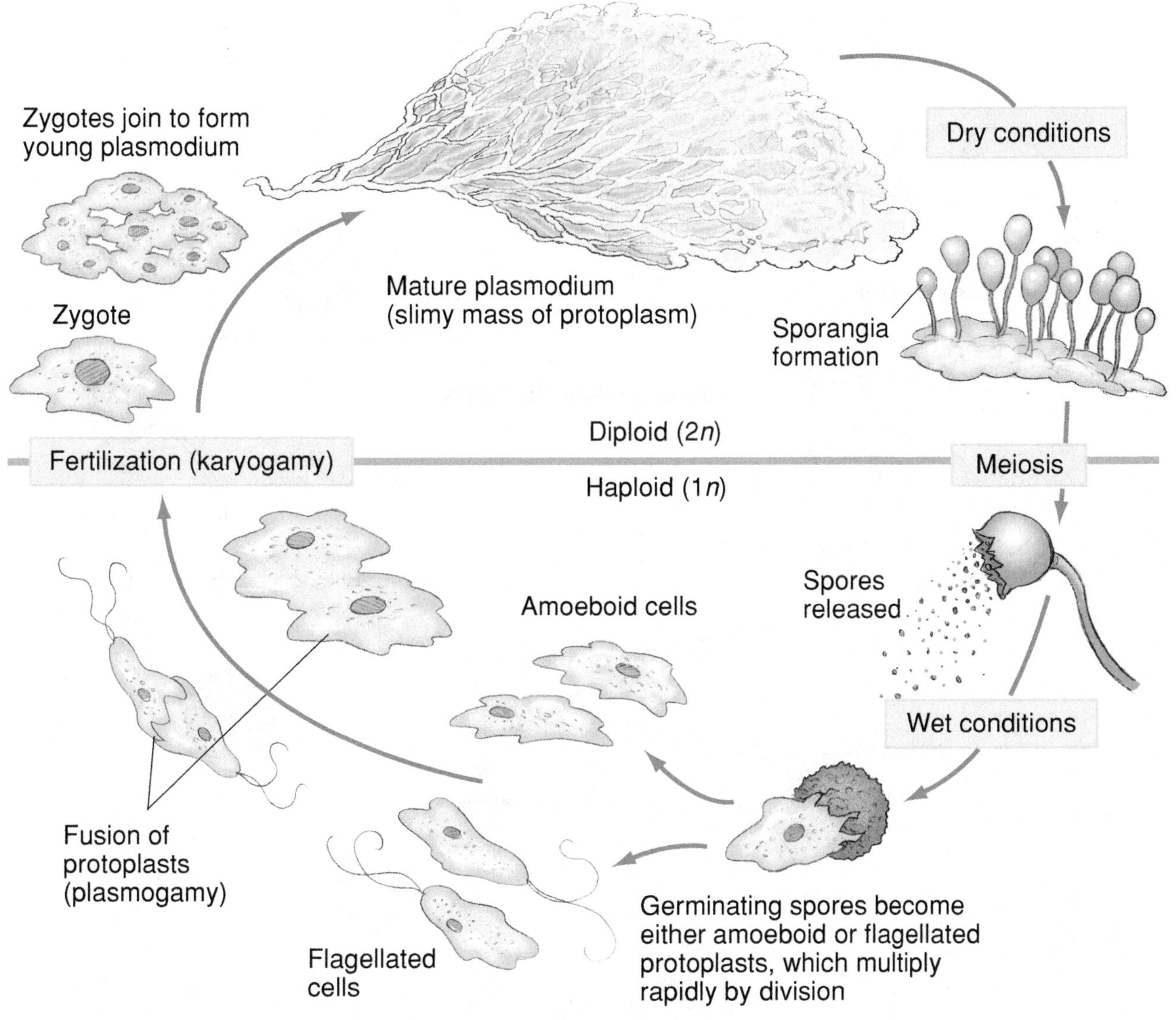

FIGURE 10.10 Slime Mold—© Kendall Hunt Publishing Company

ACTIVITY 3: Kingdom: Fungi

The Kingdom Fungi is another diverse group of about 100,000 species, ranging from unicellular to filamentous and multicellular forms. They include yeast, molds, mildews, and mushrooms, to name a few. Like plants, they have cell walls in which most contain the polymer chitin, rather than cellulose. They lack chlorophyll and are heterotrophic. They digest their food externally and then absorb it.

The body of most fungi consists of long filaments called hyphae, forming a branched network, a mycelium. Some hyphae are multinucleated and others are divided by cell walls called septa. Their life cycle may be rather complex, alternating between asexual and sexual stages. Most fungi reproduce by spores which are non motile and are transported by wind or animals. These spores can be produced asexually by mitosis or sexually by meiosis.

An example of a common fungi is *Rhizopus,* black bread mold, belonging to the Phylum Zygomycetes. Their hyphae penetrate the bread and absorb nutrients. Figure 10.11 shows its life cycle. Notice both the asexual and sexual stage. Some hyphae grow upward and produce sporangia where black asexual

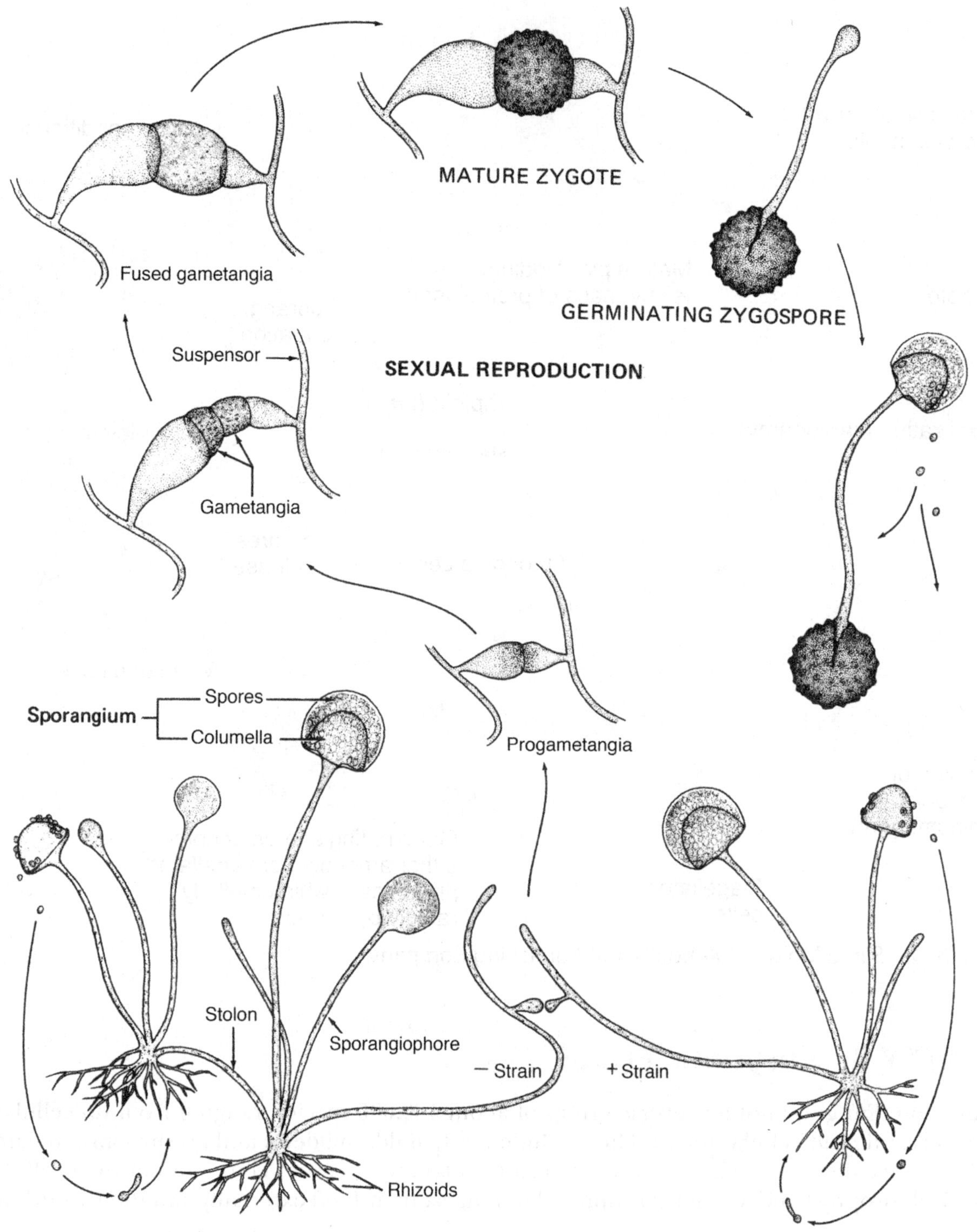

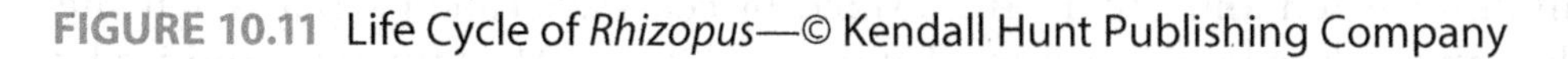

FIGURE 10.11 Life Cycle of *Rhizopus*—© Kendall Hunt Publishing Company

spores develop, giving *Rhizopus* its characteristic color. When conditions are favorable, the spores are released. Sexual reproduction occurs when spores from two different mating strains, designated + and −, meet. Their nuclei fuse to form a zygote, and an eventual thicker covered zygospore. The zygospore may remain dormant before germinating into a sporangium. This eventually releases haploid spores which develop into new hyphae, thus completing the cycle.

Another group of fungi are the sac fungi, the Phylum Ascomycetes shown in Figure 10.12. These reproduce by the production of haploid ascospores which develop in small sacs called asci. Asexually they produce projections of the hyphae called conidiophores which produce spores or conidia. Some common members include yeasts, molds, and edible forms such as truffles and morels. Only the unicellular yeast will be examined in this exercise.

Most of the fungi known commonly as the mushrooms belong to the phylum Basidiomycetes. Examples of these are seen in Figure 10.13. Their name comes from the presence of a club shaped basidium which produces basidiospores. The body of the mushroom consists of the cap supported by the stalk or stipe. On the underneath surface of the cap are the gills, thin perpendicular plates where spores are produced. Refer to Figure 10.14.

Procedure

1. Obtain a closed Petri dish with a culture of *Rhizopus,* black bread mold. KEEP THE DISH CLOSED so that the spores will not be released into the air. Examine the closed dish under a stereoscopic microscopic. Note the **hyphae, mycelia, sporangia,** and **spores**. Diagram a sample of the organism in the space provided on the Laboratory Report Sheet.
2. The yeast, *Saccharomyces cerevisiae,* will be observed. Using a Pasteur pipette, take a drop of the yeast culture that has been prepared by adding a small amount of dry yeast to warm water. Place the drop on the center of a clean microscope slide and add a coverslip. Examine under low and high power. Note the **unicellular structure.** The formation of **buds,** a form of asexual reproduction, should be observed. Diagram several cells including the bud in the space provided on the Laboratory Report Sheet.
3. Obtain a fresh or preserved specimen of a mushroom, such as *Coprinus,* and observe its structure. Some of the underground mycelium may be present. Observe the stalk or stipe supporting the upper cap. Look on the undersurface of the cap for gills. If a fresh mushroom is available, take a small section of the gill and observe it under the stereoscopic microscope. See if any spores are present. Diagram the mushroom and label the parts observed in the space provided on the Laboratory Report Sheet.

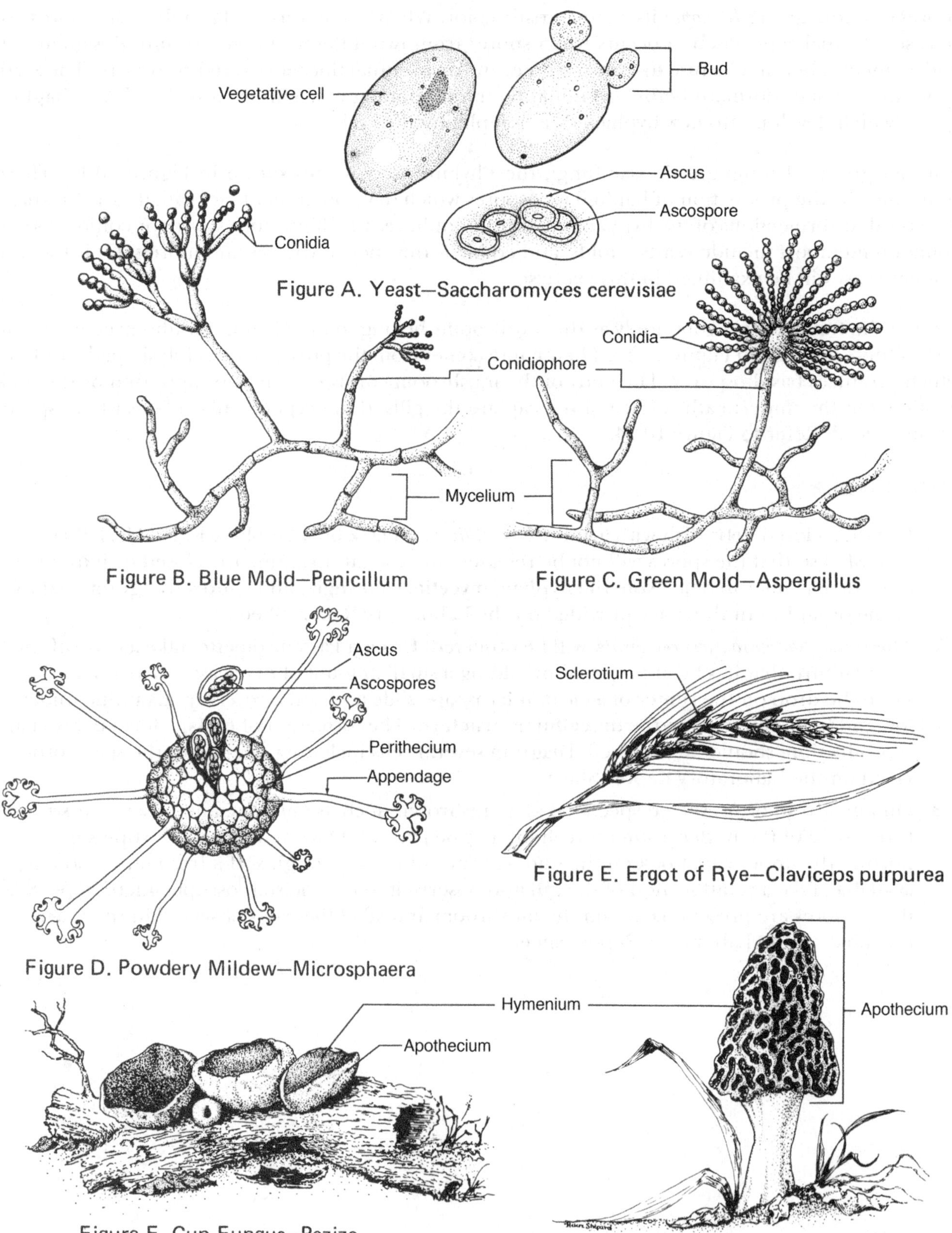

FIGURE 10.12 Types of Ascomycetes (As shown above)—© Kendall Hunt Publishing Company

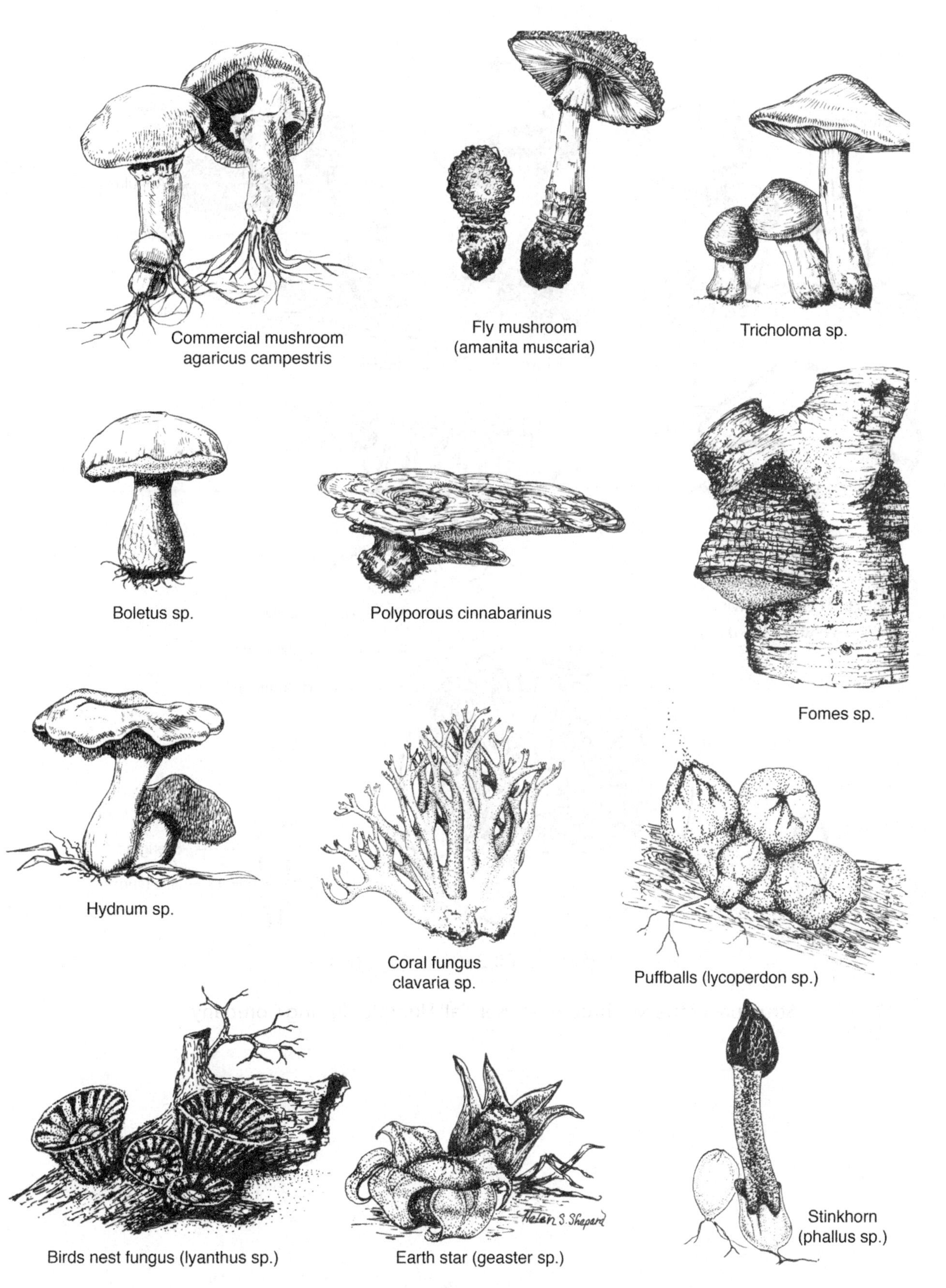

FIGURE 10.13 Types of Basidiomycetes (Higher) (As shown above)—© Kendall Hunt Publishing Company

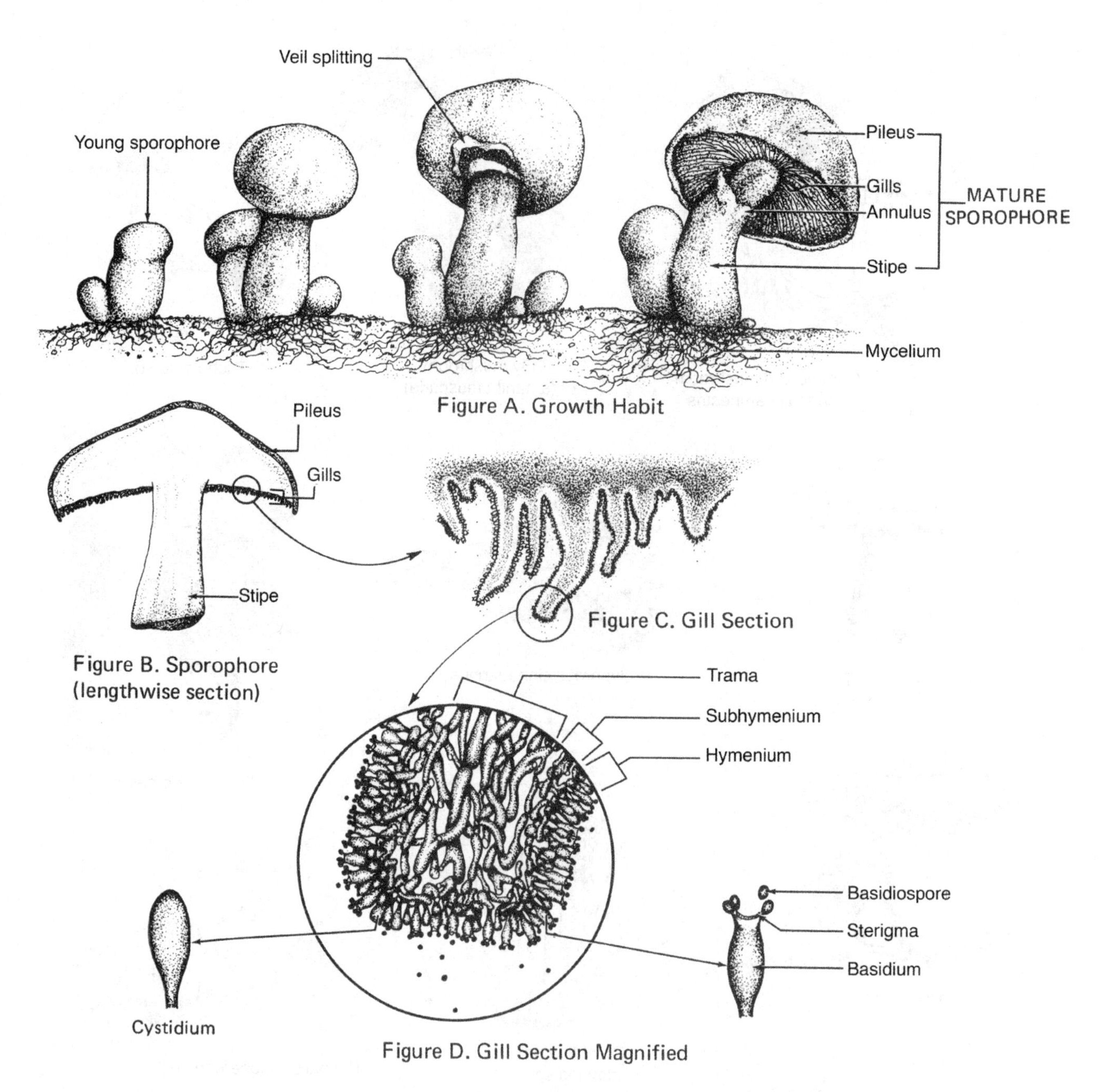

FIGURE 10.14 Structure of the Mushroom—© Kendall Hunt Publishing Company

Laboratory Report Sheet

NAME ______________________ **SECTION** __________ **GRADE** __________

Classification and Survey of the Kingdoms: Prokaryote, Protista, and Fungi

1. List the five kingdoms: ____

   ____

2. The two groups belonging to the Prokaryote are __________ and __________.

3. Diagram the three types of bacteria observed.

   ______________ ______________ ______________

4. Diagram and label *Anabaena* as observed.

5. State the function of the heterocysts: ____

6. List the three methods of locomotion of the three animal-like protists observed. Diagram and label each of them:

   ______________ ______________ ______________

7. List the phylum of the following:
   a. Diatoms: ______
   b. Green algae: ______
   c. Brown algae: ______

8. Diagram several different diatoms observed:

9. Diagram and label *Spirogyra* as observed:

10. Identify the group of Protists that are characterized by:
    a. bodies often formed of hyphae: ______
    b. photosynthetic: ______

11. Define plasmodium: ______

12. Diagram and label *Physarum* as observed:

13. Define the following:
    a. hyphae: ______
    b. mycelium: ______
    c. mating strains: ______
    d. asci: ______
    e. gills: ______

14. Diagram *Rhizopus* as it appeared under the stereoscopic microscope and label the parts observed.

15. Diagram *Saccharomyces* as observed, including the bud.

16. Diagram and label the parts of the mushroom observed.

17. State the common name of:

a. *Coprinus:* ____________________

b. *Rhizopus nigra:* ____________________

c. *Saccharomyces:* ____________________

18. State the Kingdom and Phylum or major group of the following:

	Kingdom	Phylum
a. *Rhizopus*	________	________
b. *Anabaena*	________	________
c. *Paramecium*	________	________
d. *Spirogyra*	________	________
e. *Saccharomyces*	________	________
f. Kelp	________	________
g. Mushroom	________	________

19. Identify the organism, either by genus or group name, that is associated with:

a. reproduce by budding: ____________________

b. coccus or bacillus: ____________________

c. has a cap with gills: ____________________

d. spiral-shaped chloroplast: ____________________

e. move by cilia or pseudopods: ____________________

20. Identify the Kingdom best characterized by the following:

a. multicellular; heterotrophic; motile: ____________________

b. eukaryotic; no tissue formation: ____________________

c. multicellular; autotrophic: ____________________

d. lack a definite nucleus: ____________________

e. multicellular; heterotrophic; non-motile:

____

11 GENETICS AND HUMAN VARIATION

Objectives

To be able to

1. Describe the different patterns of inheritance and apply appropriate terminology.
2. Interpret genetic data used in monohybrid crosses and use these data to predict the phenotypes of the next generation.
3. Solve genetic problems involving dominant/recessive, incomplete dominance, codominance, and sex-linked inheritance patterns.

Terms to Define

1. Gene ______________________________
2. Allele ______________________________
3. Genotype ______________________________
4. Phenotype ______________________________
5. Homozygous ______________________________
6. Heterozygous ______________________________
7. Dominant ______________________________
8. Recessive ______________________________
9. Polygenic inheritance ______________________________
10. Probability ______________________________
11. Incomplete dominance ______________________________
12. Codominance ______________________________

Chromosomes and Patterns of Inheritance

Human somatic cells (all cells except egg and sperm) contain 46 chromosomes. A **chromosome** is a single molecule of DNA with its associated proteins. These chromosomes exist as 23 pairs, called **homologous chromosomes**. The homologous chromosomes encode the same traits. One chromosome of each pair comes from the mother and one from the father at the time the sperm fertilizes the egg. The sperm and the egg each have 23 chromosomes, meaning that when they combine, they create a cell (the zygote) with 46 chromosomes.

Each chromosome may contain thousands of genes. A **gene** is a functional segment of DNA that carries the information needed to direct the synthesis of a specific protein. The homologous chromosomes contain genes for the same traits. However, the homologous chromosomes may contain different forms of the gene. These alternative forms of genes that are found at a specific locus (or location) on a chromosome are called **alleles**. Because individuals have two chromosomes for a trait, each individual has two alleles for each trait. The alleles that are present are referred to as the **genotype**. The genotype represents the primary nucleotide sequence contained in the chromosome and its component genes. The observable characteristics or traits of the individual, the **phenotype**, depend on what alleles are present. In other words, the genotype determines the phenotype. The phenotype that results depends on how the alleles interact with each other in the various patterns of heredity.

If the two chromosomes of a homologous pair carry the same allele for a particular gene, the individual is said to be **homozygous** for the trait encoded by that gene. The allele that is present will definitely be expressed in the phenotype. If the two chromosomes of a homologous pair carry different alleles for the same gene, the individual is said to be **heterozygous** for the trait encoded by that gene. The phenotype of the individual will depend on the type of interaction between the two alleles.

Alleles can be dominant or recessive. The **dominant** allele determines the phenotype, whereas the **recessive** allele is suppressed. In other words, the effects of a recessive allele will only be seen if a copy of the dominant allele is not also present. A single gene can have many different alleles, some of which are dominant and some of which are recessive.

Inheritance is the process of genetic transmission of characteristics from parents to offspring. There are several patterns of inheritance. **Simple inheritance** involves the interaction of a single pair of alleles to determine the phenotype for a particular trait. Simple inheritance patterns include dominant/recessive inheritance, codominance, and incomplete dominance. Each of these will be addressed in this laboratory exercise. **Polygenic inheritance** involves the interaction of two or more genes to determine a trait. These patterns of inheritance involve the autosomal chromosomes (chromosomes 1–22). Inheritance involving the sex chromosomes (sex-linked genes) constitutes another pattern of inheritance.

Probability

Probability problems can be used to predict the genotypes and phenotypes that will result from the mating of a particular set of parents. **Probability** is the chance that an event will happen. It is often expressed as a percentage or fraction.

$$\text{Probability} = \frac{\text{the number of events that can produce a given outcome}}{\text{the total number of possible outcomes}}$$

For example, one can determine the probability of getting a "heads" when flipping a coin. The total number of times that the "heads" can occur is once. The total number of possible outcomes equals the total number of sides on a coin, or two. Therefore, the probability of getting "heads" when flipping a coin is:

$$\text{Probability} = \frac{\text{number of events that produce a "heads"}}{\text{total number of sides on a coin}} = \frac{1}{2}$$

The probability of getting "heads" when flipping a coin is ½ or 50%.

One can also determine the probability of two independent events occurring at the same time. The probability of two independent events occurring simultaneously is equal to the product of the probability of each event occurring by itself. In other words, determine the probability for each event, and then multiply them together. For example, one can determine the probability of rolling a five on one die at the same time as rolling a two on the other die. First, determine the probability of rolling a five with one die. There are six sides on a die and the sides are numbered one through six. Therefore, the number of events that can produce a five is one. The total number of possible outcomes is six. The probability of rolling a five is, therefore, 1/6. What is the probability of rolling a two? ______________

The probability of rolling a five and a two is the product of their individual probabilities.

$$1/6 \times 1/6 = 1/36$$

There is a 1/36 chance of rolling a five and a two at the same time. Remember that a probability is not the same as definite. It is possible for you to roll two dice 36 times and roll a five and a two together once, 30 times, or not at all.

ACTIVITY 1: Probability Problem

Work in pairs. One group member will flip a coin 16 times while the other group member records the results as "heads" or "tails." Record the results in the table here.

Coin Side	Actual Result	Expected Result	Difference
Heads			
Tails			

Why is your actual result different from your expected result?

Predicting Inheritance Using Punnett Squares

A Punnett square can be used to predict inheritance when an allele can be characterized as recessive or dominant. It shows the probable offspring resulting from a **cross** (or mating) between two parent organisms. An allele is represented by a letter. Dominant alleles are represented by capital letters, whereas recessive alleles are represented by lowercase letters. The following is an example of a **monohybrid cross** that follows the inheritance of one trait.

Let's use the presence of dimples as an example. Dimples (D) are dominant over the absence of dimples (d). In the following example Punnett square, a mother who is homozygous for dimples (DD) mates with a father who is homozygous for the absence of dimples (dd). The possible mother's alleles are written across the top of the Punnett square. Each allele represents a possible gamete (egg cell), because each gamete will have only one copy of each chromosome, and, therefore, only one allele for a given trait. The father's alleles are written down the left side of the Punnett square. By combining one allele (gamete) from the mother with one from the father, you can see the genotypes of the possible offspring.

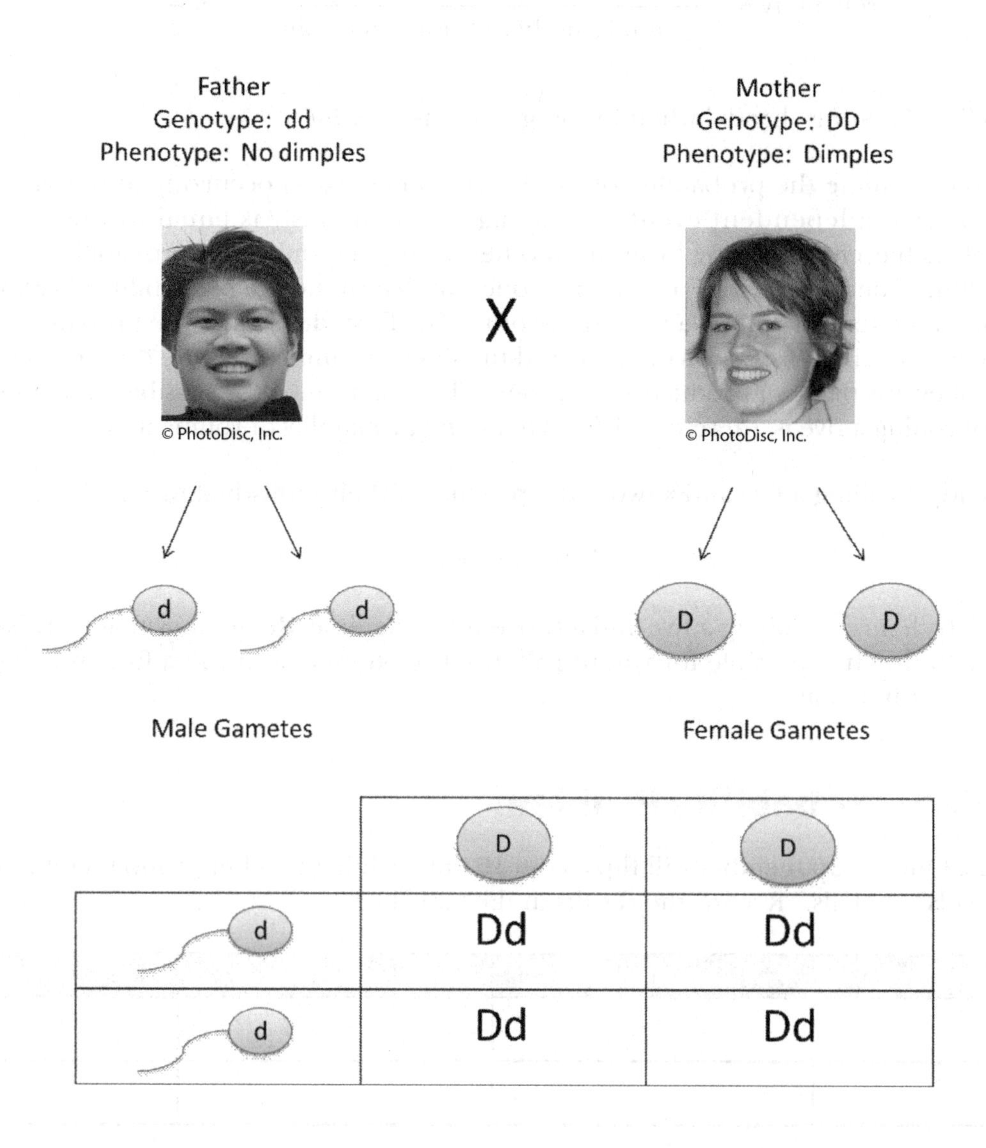

Now what does each of the genotypes mean? That depends on the type of inheritance as will be described below.

Dominant/Recessive Inheritance

Recall that in simple inheritance, one pair of alleles determines the phenotype. In dominant/recessive inheritance, the dominant allele determines the phenotype, whereas the recessive allele is suppressed. The effects of the recessive allele are only seen if a copy of the dominant allele is not also present (the individual is homozygous recessive for the allele). The effects of the dominant allele are seen even when a recessive allele is present (the individual is homozygous dominant or heterozygous).

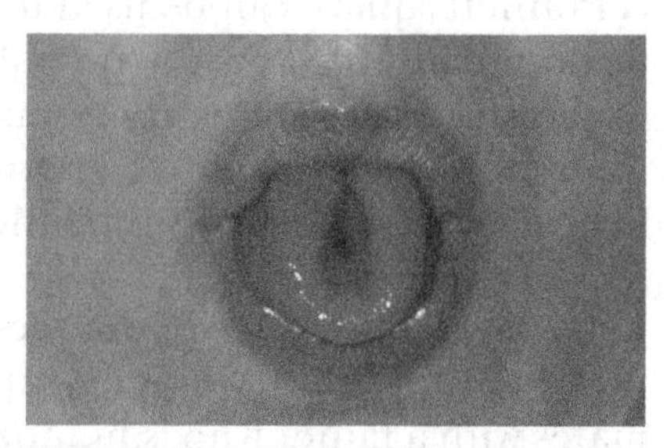

Figure 11.1 The ability to roll the tongue is a dominant trait.

The ability to roll one's tongue is inherited via dominant/recessive inheritance with tongue-rolling being dominant over the inability to roll the tongue. A woman who cannot roll her tongue marries a man who can roll his tongue, but he is heterozygous for the trait. What is the probability that they will have a child who cannot roll his/her tongue?

Use the following steps to solve the problem:

1. ***Assign a symbol to represent each type of allele.*** A capital letter (T) will represent the dominant allele (ability to roll the tongue), and a lowercase letter (t) will represent the recessive allele (inability to roll the tongue).
2. ***Determine each parent's genotype.*** The mother cannot roll her tongue, so she is homozygous recessive and has the genotype tt. The father can roll his tongue, but he is heterozygous for the trait. His genotype is Tt.
3. ***Determine the possible gametes that can be produced by each parent.*** The mother can produce only t gametes. The father can produce a T gamete or a t gamete.
4. ***Use a Punnett square to determine the possible genotypes of the offspring.***

Possible Father's Gametes	Possible Mother's Gametes: t	Possible Mother's Gametes: t
T	Tt	Tt
t	tt	tt

5. ***Determine the probability of each genotype (the genotypic ratio).*** The possible genotypes are:

   Tt: 2/4 (1/2 or 50%)
   tt: 2/4 (1/2 or 50%)

6. ***Determine the probability of each phenotype (the phenotypic ratio).*** The genotype Tt produces the phenotype with the ability to roll the tongue. The genotype tt produces the phenotype with the inability to roll the tongue. Therefore, the phenotypic ratios are:

   Ability to roll tongue: 2/4 (1/2 or 50%)
   Inability to roll tongue: 2/4 (1/2 or 50%)

7. ***Answer the question:*** What is the probability that these parents will have a child who cannot roll his/her tongue?

ACTIVITY 2: Dominant/Recessive Inheritance Problems

1. Two individuals who are heterozygous for the ability to roll the tongue mate.

   What is the father's genotype?

   What is his phenotype?

   What gametes can he produce?

What is the mother's genotype?

What is her phenotype?

What gametes can she produce?

What is the probability that they will have a child who can roll his/her tongue? (Use a Punnett square to answer the question.)

2. Albinism (a deficiency of or inability to produce the pigment melanin) is inherited in a homozygous recessive manner. A man who is homozygous dominant for normal melanin production mates with a woman who is heterozygous for normal melanin production.

   What is the man's genotype?

   What is his phenotype?

   What gametes can he produce?

   What is the woman's genotype?

   What is her phenotype?

   What gametes can she produce?

   What is the probability of them having a child with albinism?

3. Dave, who has dimples, marries Diane, who does not have dimples. Dave's father had dimples, but his mother did not. Neither of Diane's parents had dimples.

   What is Diane's genotype?

   What gametes can she produce?

   What is Dave's genotype?

   What gametes can he produce?

   What is the probability that their children will have dimples?

4. The presence of Rh factor on red blood cells (Rh positive) is dominant over the absence of Rh factor (Rh negative). What is the probability that two heterozygous parents will have a child who is Rh negative?

Incomplete Dominance

In **incomplete dominance**, heterozygous alleles produce a phenotype that is distinct from individuals who are homozygous for one allele or the other. For example, in snapdragons, the dominant allele (R) encodes red flowers, and the recessive allele (r) encodes white flowers. Plants that are homozygous dominant (RR) have red flowers. Plants that are homozygous recessive (rr) have white flowers. Plants that are heterozygous (Rr), however, have flow-ers that are not red or white. Instead, heterozygous plants have pink flowers. Therefore, the phenotype is distinct from that of either homozygous dominant or homozygous recessive plants.

If a white snapdragon is crossed with a pink snapdragon, what percentage of their offspring is likely to be red?

Use the following steps to solve the problem:

1. ***Assign a symbol to represent each type of allele.*** A capital letter (R) will represent the dominant allele, whereas a lowercase letter (r) will represent the recessive allele.
2. ***Determine each parent's genotype.*** The white plant is homozygous recessive (rr), whereas the pink plant is heterozygous (Rr).
3. ***Determine the possible gametes that can be produced by each parent.*** The white plant can produce only gametes with the r allele. The pink plant can produce an R gamete or an r gamete.
4. ***Use a Punnett square to determine the possible genotypes of the offspring.***

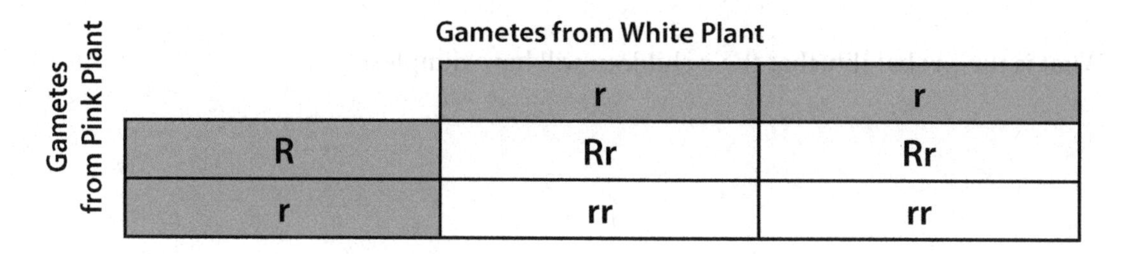

Gametes from Pink Plant \ Gametes from White Plant	r	r
R	Rr	Rr
r	rr	rr

5. ***Determine the probability of each genotype (the genotypic ratio).*** The possible genotypes are:

Rr: 2/4 (1/2 or 50%)
rr: 2/4 (1/2 or 50%)

6. ***Determine the probability of each phenotype (the phenotypic ratio).*** The genotype RR produces the phenotype with red flowers. The genotype Rr produces the phenotype with pink flowers. The genotype rr produces the phenotype with white flowers. Therefore, the phenotypic ratios are:

Red: 0/4 (0%)
Pink: 2/4 (1/2 or 50%)
White: 2/4 (1/2 or 50%)

7. ***Answer the question:*** What is the probability that these plants will have offspring with red flowers?

ACTIVITY 3: Incomplete Dominance Problems

Red blood cells (RBCs) transport oxygen using the protein hemoglobin. The allele A encodes the normal hemoglobin protein. Individuals with the genotype AA will have normal hemoglobin and normal RBCs. The allele S encodes the abnormal hemoglobin protein. Individuals with the genotype SS produce only the abnormal hemoglobin protein, and their red blood cells take on a sickled (crescent) shape. These individuals will have sickle cell anemia. The sickled RBCs clump together and cannot transport oxygen adequately. The RBCs may clog the capillaries, causing inflammation, pain, and organ damage. Individuals with the AS genotype produce both the normal and abnormal forms of hemoglobin. They are said to have the sickle cell trait and have RBCs that only sickle in extremely low oxygen environments.

Malaria is a parasitic disease that is transmitted by mosquitoes and is common in tropical areas. It lives in red blood cells. The presence of the parasite in the RBCs of an individual with the abnormal hemoglobin allele (S) will cause the RBCs to sickle, thereby causing the death of the parasite. Individuals with the SS or AS genotypes are resistant to malaria, whereas those with the AA genotype are susceptible.

1. Jane and John both have the sickle cell trait. What is the probability that their children will have sickle cell anemia?
2. What is the probability that Jane and John's children will be resistant to malaria?

Codominance

In **codominance**, neither allele is dominant over the other. The phenotype of an individual who is heterozygous is a combination of both fully expressed traits. An example is coat color in cattle. Cattle with red coats are homozygous (R^1R^1). Cattle with white coats are also homozygous, but for a different allele than red coated cattle. White cattle have the genotype R^2R^2. Cattle that are heterozygous (R^1R^2), exhibit a combination of the red and white coat traits. They have patchy red and white coats (termed roan).

Two roan cattle mate. Is it possible for them to have red offspring?

Use the following steps to complete the problem:

1. ***Assign a symbol to represent each type of allele.*** R^1 will represent the red allele. R^2 will represent the white allele.
2. ***Determine each parent's genotype.*** Both parents are roan, so their genotypes are R^1R^2.
3. ***Determine the possible gametes that can be produced by each parent.*** Each parent can produce an R^1 gamete and an R^2 gamete.
4. ***Use a Punnett square to determine the possible genotypes of the offspring.***

Male Gametes \ Female Gametes	R^1	R^2
R^1	R^1R^1	R^1R^2
R^2	R^1R^2	R^2R^2

5. ***Determine the probability of each genotype (the genotypic ratio).*** The possible genotypes are:

R^1R^1: 1/4 (25%)
R^1R^2: 2/4 (1/2 or 50%)
R^2R^2: 1/4 (25%)

6. ***Determine the probability of each phenotype (the phenotypic ratio).*** The genotype R^1R^1 produces red cattle, the R^1R^2 genotype produces roan cattle, and the R^2R^2 genotype produces white cattle. Therefore, the phenotypic ratios are:

   Red: 1/4 (25%)
   Roan: 2/4 (1/2 or 50%)
   White: 1/4 (25%)

7. ***Answer the question:*** Is it possible for them to have red offspring?

ACTIVITY 4: Codominance

ABO blood type is inherited via codominance and multiple allelism (i.e., there are more than two alleles for ABO blood type in the human population). The ABO blood system has three alleles of the blood group gene that encodes enzymes that synthesize sugars (antigens) found on the surface of RBCs: I^A (or A), I^B (or B), and i (or O). I^A and I^B are codominant, whereas i is recessive. People make antibodies against the sugars (antigens) that they do not have on their cells. Table 11.1 summarizes possible genotypes and phenotypes under the ABO blood system.

Table 11.1 The Genetics of ABO Blood Typing

Blood Type	Red Blood Cell Antigens	Plasma Antibodies
Type A (AA, AO)	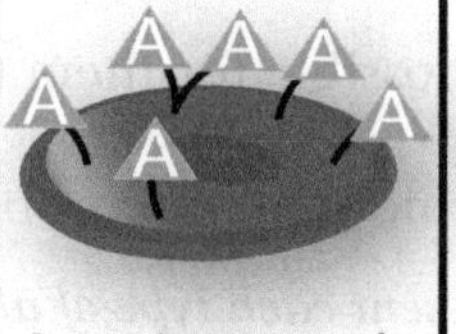 A antigens only	 antibody b only
Type B (BB, BO)	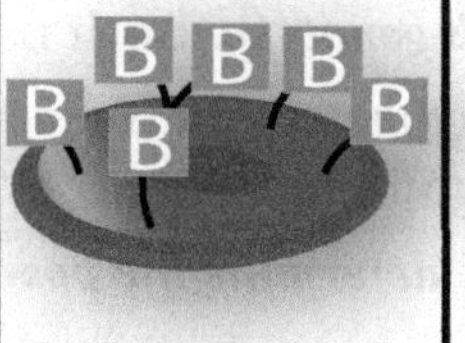 B antigens only	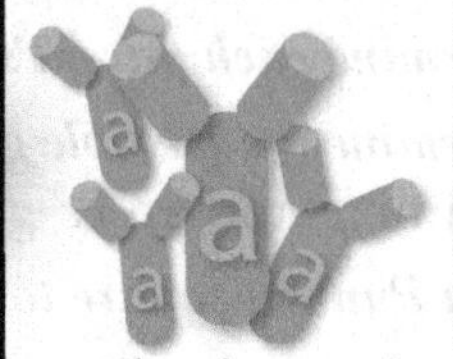 antibody a only
Type AB (AB)	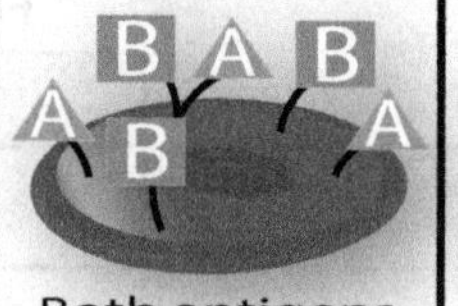 Both antigens A and B	None Neither antibodies a nor b
Type O (OO)	 Neither antigens A nor B	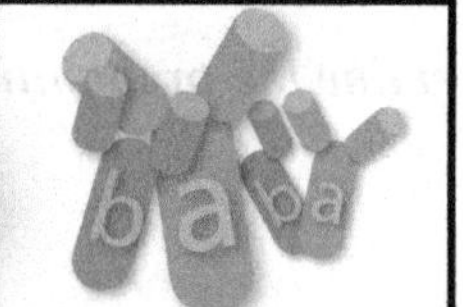 Both antibodies a and b

Courtesy of Catherine Rappazzo. © Kendall Hunt Publishing Company

Determining the ABO blood type can be used to help establish identity and paternity and is important in blood transfusions. If a person receives blood that is incompatible to theirs (i.e., against which they have antibodies), a transfusion reaction may occur. The antibodies in the recipient's blood attack the sugars on the RBCs in the donor blood, causing the RBCs to clump together. This clumping can block blood vessels. Hemolysis (breakdown of red blood cells) may also occur. Transfusion reactions may be fatal.

1. A man with type AB blood marries a woman with type A blood. Is it possible for them to have a child with type B blood? Use Punnett squares to support your answer.

2. A man with type O blood and a woman with type AB blood marry. What is the probability that they will have a child with type A blood?

Sex-Linked Inheritance

Genes located on the X or Y chromosome are called sex-linked genes. Y-linked genes are located on the Y chromosome and passed from father to son. Females inherit X-linked genes from the mother and father, whereas males only inherit X-linked genes from their mothers.

Many X-linked genes are inherited in a dominant/recessive manner. Males will always express the X-linked gene regardless of whether it is dominant or recessive, because they only have one copy of the gene. Females may be homozygous dominant, homozygous recessive, or heterozygous for X-linked genes.

Red-green color blindness is inherited in an X-linked manner. The dominant allele on the X chromosome results in normal color vision, whereas the recessive allele results in red-green color blindness. What is the probability that Sean, with normal color vision, and Julie, a carrier of color blindness, will have a son who has red-green color blindness?

Use the following steps to solve the problem:

1. ***Assign a symbol to represent each type of allele.*** X-linked alleles are written as superscripts on an X. Let's use an N to represent the dominant allele (normal color vision) and an n to represent the recessive allele (red-green color blindness). Therefore, the dominant allele is written as X^N, whereas the recessive allele is written as X^n.
2. ***Determine each parent's genotype.*** A man has one X and one Y chromosome. Sean has normal color vision, so his genotype is X^NY. Note that he has one X chromosome and one Y chromosome. Julie is a carrier of red-green color blindness. This means that she has one allele for color blindness but isn't color-blind herself. Therefore, her genotype is X^NX^n.
3. ***Determine the possible gametes that can be produced by each parent.*** Sean can produce an X^N gamete or a Y gamete. Julie can produce an X^N gamete or an X^n gamete.
4. ***Use a Punnett square to determine the possible genotypes of the offspring.*** If Sean passes on a Y allele, the offspring will be a son. If Sean passes on an X allele, the offspring will be a daughter. Sean can only pass on an allele for color vision/color blindness to his daughters.

Julie's Gametes

Sean's Gametes	X^N	X^n
X^N	$X^N X^N$	$X^N X^n$
Y	$X^N Y$	$X^N Y$

5. ***Determine the probability of each genotype (the genotypic ratio).*** The possible genotypes are:

X^NX^N: 1/4 (25%)
X^NX^n: 1/4 (25%)
X^NY: 1/4 (25%)
X^nY: 1/4 (25%)

6. ***Determine the probability of each phenotype (the phenotypic ratio).***

Daughters with normal color vision: 2/4 (1/2 or 50%). Note: Both daughters have normal vision (X^NX^N and X^NX^n), but one daughter (X^NX^n) is a carrier of red-green color blindness.

Daughters with red-green color blindness: 0/4 (0%)

Sons with normal color vision: 1/4 (25%). This is the son with the X^NY genotype.

Sons with red-green color blindness: 1/4 (25%). This is the son with the X^nY genotype.

7. ***Answer the question:*** What is the probability that Sean, with normal color vision, and Julie, a carrier of color blindness, will have a son who has red-green color blindness? ____________________

ACTIVITY 5: X-linked Inheritance

1. The disorder hemophilia A results from the inability to produce the normal blood clotting factor VIII, causing abnormal, excessive bleeding. Hemophilia A is inherited via X-linked inheritance, with the gene for normal clotting factor VIII production (XH) being dominant over the gene for abnormal production of clotting factor VIII (Xh). Edward, a hemophiliac, marries Anna who has normal blood clotting. Anna's father was also a hemophiliac. What is the probability that they will have a daughter with hemophilia?

2. Andrew has red-green color blindness. He marries Susan who is also color-blind. From which parent or parents did he inherit the color blindness allele?

Human Variation

The human genome is about 99.9% identical between any two randomly chosen people. Why do we look and act so different then? The differences in our phenotypes come from that 0.1% that is different between any two people.

Four processes contribute to genetic variation (the differences in our phenotypes): mutations, synapsis and crossing-over, random alignment, and random fertilization. Mutations are permanent changes in a DNA sequence. Repeated mutations create genetic variation in a population. Some mutations are detrimental; others provide survival advantages. Crossing-over during prophase I of meiosis I creates unique combinations of genes on chromosomes. Therefore, our chromosomes are not identical to those of our mothers or fathers. Random alignment and independent assortment occur during meiosis I. Independent assortment of 23 pairs of homologous chromosomes means that each human can make 8 million different types of gametes. Finally, which gametes (egg and sperm) combine during fertilization is a random process. The odds of receiving a particular combination of chromosomes are 1 in 8 million times 1 in 8 million or 1 in 64 trillion! That means that there is about a 1 in 64 trillion chance of you having identical genotypes and phenotypes as someone else.

ACTIVITY 6: Human Variation

In this activity, you will examine your phenotype for several easily distinguished traits. You will use the phenotype to determine your genotype. If you possess the recessive trait, you will know that you have the homozygous reces-sive genotype for the trait. If you possess the dominant phenotype, you will only be able to determine that you have at least one copy of the dominant allele. You will not be able to determine if you are homozygous dominant or heterozygous. You will write your genotype as a capital letter followed by a dash (e.g., A–), meaning you have at least one dominant allele but do not know the second allele. After you identify your genotypes and phenotypes, you will compare your data to that of the class..

Complete the following table. Descriptions of the various traits are provided below.

Trait	My Phenotype	My Genotype	Number of Students with at Least One Dominant Allele	Number of Students with No Dominant Alleles
Interlocking fingers				
PTC tasting				
Sodium benzoate tasting				
Dimples				
Tongue rolling				
Attached earlobes				
Widow's peak				
Double-jointed thumb				
Freckles				
Darwin's earpoint				
Rh blood type				
ABO blood type			n/a	n/a

- Interlocking fingers: Clasp your hands together by interlocking your fingers as shown in Figure 11.2. If your left thumb is uppermost, you possess a dominant allele (F) for the trait. If your right thumb is on top, you are homozygous recessive (ff) for the trait.

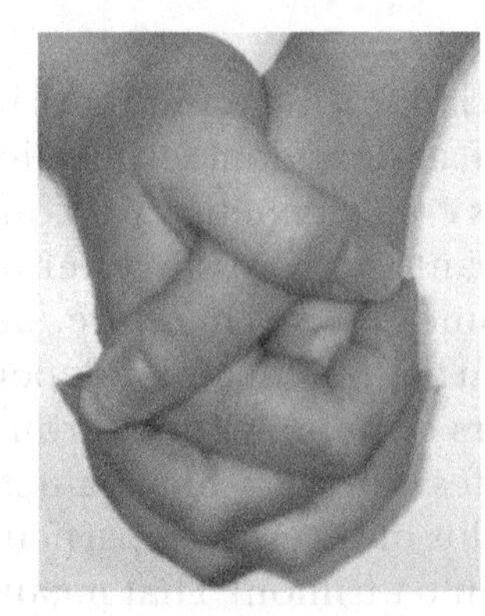

Figure 11.2. Interlocking fingers with the right thumb uppermost is a recessive trait.

- PTC tasting: PTC (phenylthiocarbamide) is a harmless chemical. It produces a bitter taste for some people, but not for others. Put the PTC strip on your tongue. If you taste the strip (anything besides a "paper" taste), you are a PTC taster and possess a dominant allele (P). Non-tasters are homozygous recessive (pp).
- Sodium benzoate: Put the sodium benzoate strip on your tongue. Tasting (S) is dominant over non-tasting (s).
- Dimples: The presence of dimples on one or both cheeks (D) is dominant over the absence of dimples (d).
- Tongue rolling: The ability to roll the tongue (see Figure 11.1) is dominant (T) over the inability to roll the tongue (t).
- Attached earlobes: Have your lab partner examine your earlobes. If your earlobe hangs free below its point of attachment to your head, you possess a dominant allele (A) for this trait. If the earlobe is attached, you are homozygous recessive (a) for the trait. (See Figure 11.3.)

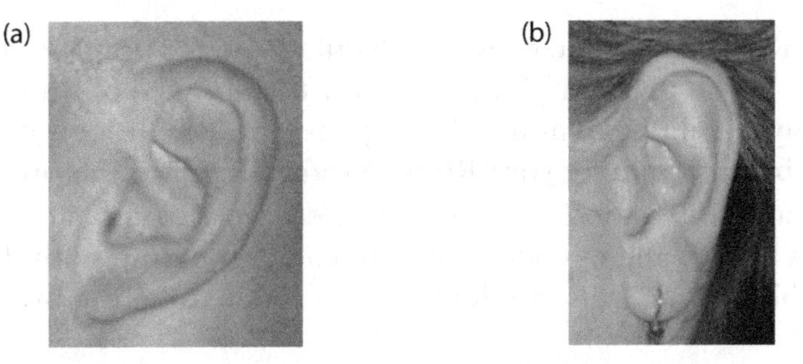

Figure 11.3 (a) Attached earlobes. (b) Free earlobes.

- Widow's peak: The widow's peak (W) is dominant over a continuous hairline (w) (see Figure 11.4).

(a)

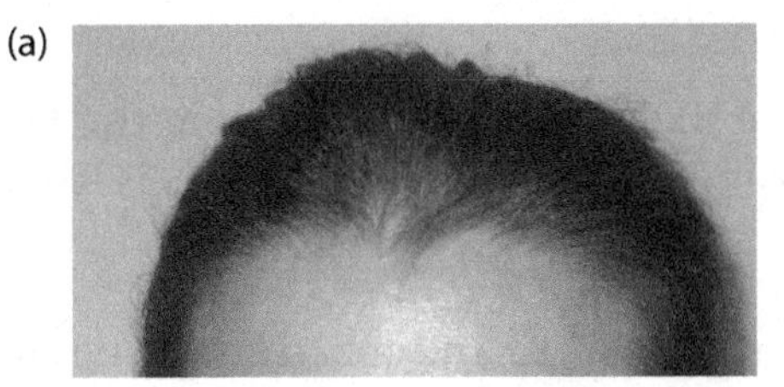

(b)

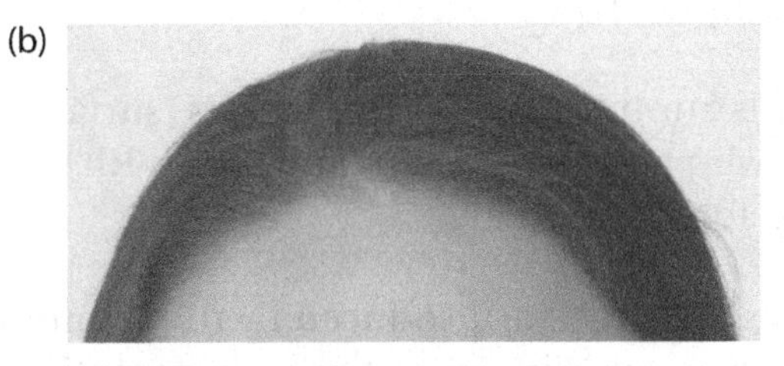

Figure 11.4 Hairlines. (a) Widow's peak. (b) Continuous hairline.

- Double-jointed (hitchhiker's) thumb: The double-jointed thumb (J) is dominant over a straight thumb (j) (see Figure 11.5).

(a)

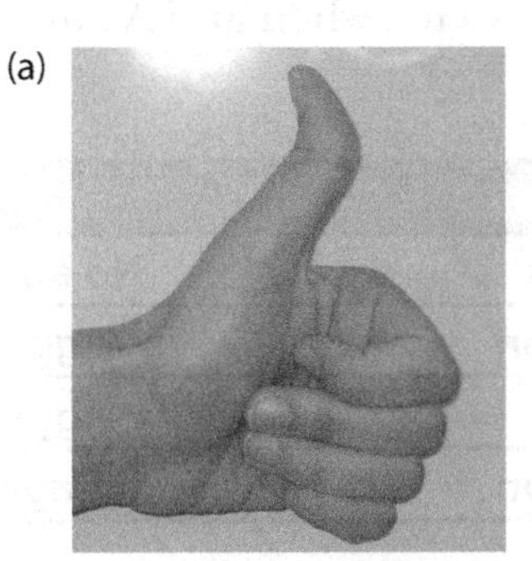

(b)

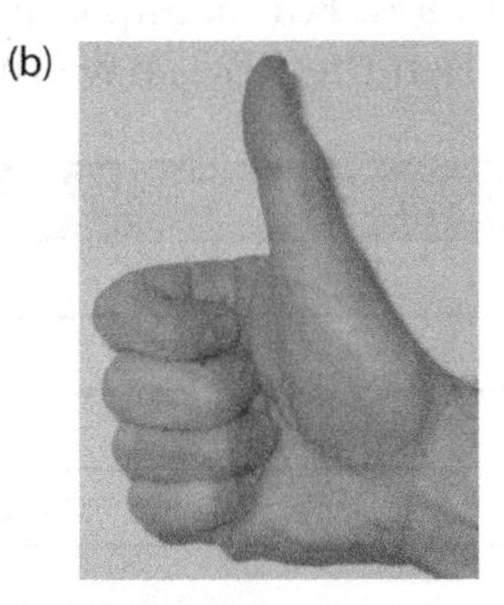

Figure 11.5 (a) Double-jointed thumb. (b) Straight thumb.

- Freckles: The presence of freckles (F) is dominant over the absence of freckles (f).
- Darwin's earpoint: Darwin's earpoint (E) is dominant over a smooth ear (e) (see Figure 11.6).

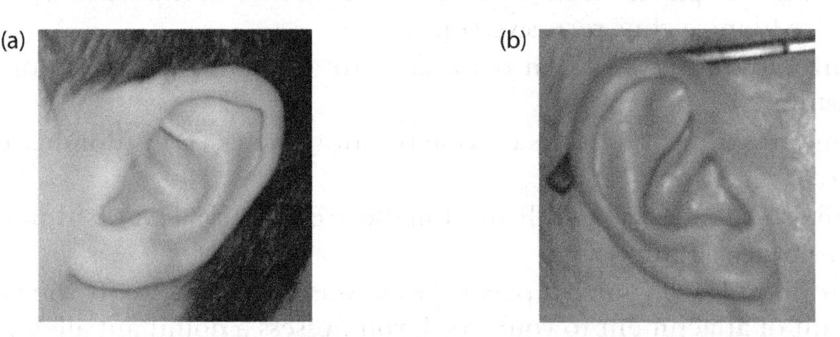

Figure 11.6 (a) Darwin's earpoint. (b) Smooth ear.

- Rh blood type: Rh (Rhesus) factor is a molecule (antigen) on the surface of red blood cells. The presence of Rh factor (+) is dominant over the absence of Rh factor (–). Individuals with the Rh– phenotype will make antibodies against the Rh antigen if they are exposed to Rh+ blood. You will be determining your Rh type today. Refer to the procedure below.
- ABO blood type: ABO blood type is inherited via incomplete dominance and multiple allelism, with IA and IB alleles codominant to each other and both dominant over i. If you do not know your blood type, you can determine it today using the procedure below.

ACTIVITY 7: Blood Typing

Blood type is determined by the presence or absence of specific antigens on the surface of red blood cells (RBCs). In the ABO blood group system, the antigens are glycoproteins (A or B antigens) on the surface of RBCs. As you saw in Table 11.1, the presence or absence of these antigens determines your ABO blood type.

Rh factor is another antigen on the RBC surface. There are six Rh antigens, including the D antigen. A person who possesses the Rh antigens is Rh+, whereas someone lacking the antigens is Rh–. We will be testing for Rh D today.

Antibodies are proteins produced by the immune system that bind to and inactivate antigens. When an antibody reacts with an antigen, an agglutination reaction occurs. Agglutination results in visible clumping of the antigen and antibody. Therefore, antibodies against these different antigens can be used for blood typing purposes.

Antiserum containing anti-A, anti-B, and anti-D antibodies will be used to perform ABO and Rh blood typing. When antiserum is mixed with blood, an agglutination reaction will occur if the specific antigen is present in the blood. For example, if agglutination occurs when anti-A antibodies are added to the blood specimen, then the RBCs have the A antigen.

Blood Type	Anti-A Antibodies	Anti-B Antibodies
A	Agglutination	No agglutination
B	No agglutination	Agglutination
AB	Agglutination	Agglutination
O	No agglutination	No agglutination

If agglutination occurs when anti-D antibodies are added to the blood, then the blood sample is Rh^+. If no agglutination occurs, the blood is Rh^-.

Procedure

1. Obtain your blood sample.
   a. Choose a finger from which you will draw the sample.
   b. Wipe the finger with an alcohol prep pad, and allow the finger to dry.
   c. Remove the cap of the automatic lancet, and place it on the fingertip (slightly off center).
   d. Push the lancet to stick your finger.
   e. Place the lancet in sharps container.
2. Place one drop of blood in each of the A, B, D/Rh wells on the card.
3. Add one drop of the anti-A antiserum to the A well.
4. Add one drop of the anti-B antiserum to the B well.
5. Add one drop of the anti-D antiserum to the D/Rh well.
6. Use separate toothpicks to stir each well.
7. Pick up the plate, and rock it back and forth to look for agglutination in each of the wells.
8. Record your blood type.
9. Throw away the toothpicks in the sharps container. Place the blood typing cards in the biohazard container.

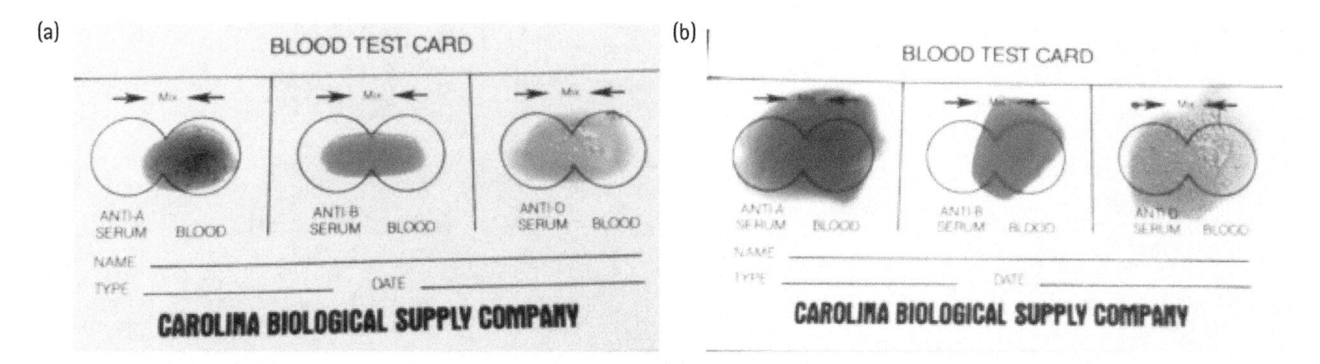

FIGURE 11.7 (a) Type A+ blood. Note the agglutination that occurs when the anti-A antibody binds to antigen A in type A blood. Note the agglutination that occurs when the anti-D antibody binds to the D antigen in Rh+ blood. (b) Type O+ blood. Note that no agglutination occurs in type O blood because there are no A or B antigens present. Agglutination with anti-D antibody indicates Rh+ blood.

Laboratory Report Sheet

NAME__________________________________ **SECTION** ____________ **GRADE** ____________

Genetics and Human Variation Laboratory Report

1. What is the probability that two individuals will have five sons in a row?

2. In humans, freckles are dominant over no freckles. A man with freckles marries a woman with no freckles. They have three children, two of whom have freckles and one of whom does not.
   a. What is the man's genotype?

   b. What kind(s) of gametes can he produce?

   c. What is the woman's genotype?

   d. What kind(s) of gametes can the woman produce?

   e. What are the genotypes of the children with freckles?

   f. What is the genotype of the child without freckles?

3. Free earlobes are dominant over attached earlobes. Chad, who has attached earlobes, marries Andrea, who has attached earlobes. Is it possible for them to have a child with free earlobes? Why or why not?

4. Cystic fibrosis is one of the most common autosomal recessive diseases in people of Northern European descent. A mutation in the CF gene affects a protein involved in the transport of chloride and sodium across cell membranes, causing thick mucus and secretions, lung damage, and nutritional deficiencies. The disease is inherited in an autosomal recessive manner, meaning one must inherit two mutated alleles in order to have the disease. Greg does not have cystic fibrosis, and there is no family history of the disease. His wife Sally also does not have cystic fibrosis, but her sister does. As a genetic counselor, what would you tell Greg and Sally about the probability of their children inheriting cystic fibrosis? In your explanation, include the possible genotypes for both Greg and Sally.

5. Sickle cell anemia is inherited via incomplete dominance. The allele A encodes the normal hemoglobin protein. Individuals with the genotype AA will have normal hemoglobin. The allele S encodes the abnormal hemoglobin protein. Individuals with the genotype SS produce only the abnormal hemoglobin protein and have sickle cell anemia. Individuals with the AS genotype produce both the normal and abnormal forms of hemoglobin. They have the sickle cell trait and have RBCs that only sickle in extremely low oxygen environments. The presence of the sickle cell hemoglobin allele (S) also provides resistance to malaria.

   A man and woman living in a tropical area where malaria is prevalent and health care is inaccessible have five children. Their genotypes are SS, AS, AS, SS, and AA.

   a. What are the genotypes of the parents?

   b. What kinds of gametes can the mother produce?

   c. What kinds of gametes can the father produce?

   d. Why would children who are AS be most likely to live to adulthood and reproduce?

6. ABO blood type is inherited via multiple allelism and codominance. A woman with type A blood and a man with type B blood get married. Is it possible for these individuals to have a child with type O blood?

7. The disorder hemophilia A is inherited via X-linked inheritance with the gene for normal clotting factor VIII production (X^H) being dominant over the gene for abnormal production of clotting factor VIII (X^h). A woman who is a carrier for hemophilia A marries a man with normal blood clotting.
   a. What is the genotype of the woman?

   b. What is the genotype of the man?

   c. What is the probability of them having each of the following children:

      i. A girl with normal blood clotting?

      ii. A girl with hemophilia?

      iii. A boy with normal blood clotting?

      iv. A boy with hemophilia?

8. Kathy has type A^+ blood. Is it safe for her to donate blood to Michael who has blood type O^+? Why or why not?

9. What are the four factors that contribute to genetic variation?

10. Were any of the traits that you studied expressed in an identical way by all of the members of your class? Why is it unlikely for all the individuals in a class to express the same phenotype for all traits?